D1053364

SEEDS

The Definitive Guide to Growing, History, and Lore

The Definitive Guide to Growing, History, and Lore

PETER LOEWER

Macmillan • USA

MACMILLAN
A Simon & Schuster Macmillan Company
1633 Broadway
New York, NY 10019

Library of Congress Cataloging-in-Publication Data
Loewer, H. Peter.
Seeds : The definitive guide to growing, history, and lore/Peter Loewer.
p. cm.
Includes bibliographical references and index.
ISBN 0–02–574042–3
1. Seeds. 2. Seed technology. 3. Seed industry and trade.
4. Plant propagation. I. Title
SB117.L59 1996
631.5'21—dc20 95-35876 CIP

Printed in the United States of America
Book Design: Rachael McBrearty

10 9 8 7 6 5 4 3 2 1

ACKNOWLEDGMENTS

A number of gardeners, seedsmen, and seedswomen have helped me gather the information for this book and most of them are mentioned in the text. But there is one person in the world of horticulture that knows more about seeds than anyone else I have ever met, and his name is Magella Larochelle, now of Quebec, Canada. I'd like to thank him here for sharing so much of his knowledge. And thanks, too, to Pam Hoenig of Macmillan; Dominick Abel, my agent; Peter and Jasmin Gentling of Asheville; and my wife Jean.

CONTENTS

INTRODUCTION

All About Seeds

Seeds are earthbound starships that fly or fall through the air, never leaving our planet. Both seed and ship must provide a wall of protection for life within and enough food and oxygen to support that life until the ship or seed opens to a brave new world. If seeds were carried across space to distant planets and the conditions they found were suitable to their internal workings, they would sprout.

Seeds come as large as a coconut or as small as the dust motes produced by orchids or begonias. Seeds are round, curved, feathered, pitted, surrounded by fruit, or loose in a pod. Like the silver maple seed of our northeastern forests, some die within a few days of falling to earth, while many survive for years waiting for just the right combination of moisture, heat, and time to force germination. A few, like the regal lotus, may wait within a tomb for centuries (the lotus seed was once buried as an offering to eternity by a long-departed culture) but still be able to spring to life after about 1,040 years. In the interest of accuracy, however, chapter 2 talks about a seed's age and viability.

Seeds provide more food for the human race than any other plant or animal. We consume vast amounts of peanuts, rice and corn kernels, wheat, barley, oats, even pecans and pistachios. About one-fourth of the human energy demands of the United States are provided by cereal seeds, and seeds account for nearly three-quarters of the human diet in the rest of the world. We mash, grind, boil, and compress seeds into margarine, medicine, cosmetics, and alcoholic beverages, not to mention those seeds used by gardeners to serve as an incomparable source of beauty in and about the home.

Seeds are not only food, but also international symbols that have colored our languages from time immemorial. We speak of seed money, good and bad seed, and the seed of an idea. In physiological terms, seed is semen, sperm, milt, or spat. We speak of descendents as, for example, the seed of David. Seeds are the initial source for many things, both real and symbolic, like the seeds of virtue or vice. The eggs of insects, the silkworm moth to name one, are called seeds, and fishermen at sea speak of young oysters used for transplanting as seed oysters.

According to Eric Partridge in *Origins, A Short Etymological Dictionary of Modern English* (Macmillan, New York, 1958), the word *seed* is a noun or a verb, whence, *seedling* and *seedy*; *secular* (an adjective and noun), and possibly influenced by French, *secularism, secularist, secularize,* and *secularization.*

"To seed" derives from the Old English *saedian,* itself from the Old English *saed,* whence, in Middle English *sed* or *seed,* and today's word *seed.*

In 1849, a youth was called a seed. By 1900, the word *seed* referred to a young man who would never grow into anything. During that same period, a poker chip was also called a seed. Around 1930, a dollar was called a seed, and we still have *birdseed, hayseed,* and *swampseed.* Birdseed is any dry or packaged breakfast cereal; hayseed describes a farmer or a rustic and even today is used to describe hayseed theaters; swampseed refers to a dialect word for rice, especially favored by the armed forces during World War II.

Seedtime is the proper time for sowing, and a *seeder* is the one who sows the seed. And the word *seedy* can mean something that abounds with seeds; a beverage that has a particular flavor, usually due to seeds growing among the vines (once said of French brandies); or something like clothing that has become shabby, miserable, or wretched. At one time, manufactured glass that contained small bubbles was called seedy.

The hazards that face a space traveler on an alien planet or on many parts of our own earth could not be more formidable than those facing a new seed when it's thrust upon the world, even before germination. The weather brings harsh winds, snow, rain, and hail, the dry heat of a desert sun, or the rigid cold of a mountain winter. The rumbling stomach of a bird or beast and the endless enemies that creep, walk, and crawl within and on the earth's skin also decrease a seed's chances. Then, when the seedling emerges from the seed's protective husk and it's growing, these threats become even more daunting to the burgeoning life of a plant.

Since first writing the book, some of the subjects in the following chapters have changed because scientific investigation into the life processes of the seed continues with greater zeal every year. And, unfortunately, seed houses come and go, for they, like everything else these days, are not immune to the swings of the economic pendulum.

In attempting to cover a great deal about seeds, I'm well aware that many things are missing. So if you, the reader, spot such gaps or know about a great source of seeds, please write me (no e-mail, please) at Macmillan • USA, and I will answer—I always answer my mail.

CHAPTER 1

The Great Plant Kingdom

The plant kingdom ranks second to the animal kingdom in the diversity of its members, and, unfortunately, to most eyes, plants appear to have a much simpler structure than animals. But make no mistake; plants are far more complicated than many people would like to believe. Because plants do not march across the landscape like a human army does not mean they are to be scorned for lacking sophistication. Remember that plants have evolved hundreds of chemicals to defend themselves against other plants and many animals, including man. Today, most scientists realize that if plant species were classified by the chemicals they produce, the diversity in the plant kingdom would explode. And if the plant's ability to manufacture its own food through photosynthesis would suddenly cease, all life on earth would soon perish.

THE CHARACTERISTICS OF PLANT AND ANIMAL CELLS

The basis of all life is the individual cell. Living organisms begin with one-celled animals and march up the scale to complex plants and animals, each containing trillions of cells. Cells are found in several shapes and sizes, including circles, squares, ovals, and rectangles. And some animal cells actually have the ability to change their shapes. But all these cells, whether plant or animal, exhibit the following features: They are each surrounded by a *plasma* (or cell) *membrane.* Each cell holds specialized parts, called *organelles,* structures designed for a specific purpose in the cell, for example, a *nucleus* or control center. And they all contain *cytoplasm,* or the fluid portion of the cell. Taken together, these three structures are termed *protoplasm*, a term used to represent the overall contents of a living cell.

But the similarities do not end here. When a generalized animal cell is compared with a generalized plant cell, all share most of the same features.

Both types of cell will have a *plasma membrane*, but some plant cells will possess secondary cell walls that contain *cellulose*. This special polysaccharide is the most abundant of its kind, and its strength is used to shore up stems to bear additional weight.

Both plant and animal cells have a *nucleus* that contains *chromosomes* made of DNA and other proteins. Different species have different chromosome counts: Humans have forty-six, bullfrogs have twenty-six, apples have thirty-four, and garden peas have fourteen.

Both types of cells have an *endoplasmic reticulum* (ER), an elaborate and complicated system of folded membranes that extends throughout most of a cell. Endoplasmic reticulum resembles hundreds of layers of tissue paper, with space between each layer, folded, so it would completely fill a box or drawer. Endoplasmic reticulum functions as a circulatory system for the cell.

Both types of cells contain *ribosomes*, round subunits of a cell that are produced by the nucleus. Ribosomes are responsible for protein synthesis. They are found lining some of the pathways of ER and also float freely within the cytoplasm.

Both contain *mitochondria*, bean-shaped structures that contain an inner membrane that also exhibits elaborate folds. These organelles are responsible for metabolism within the cell.

Only plant cells contain the next structure, known as a *plastid*. Plastids are bean-shaped structures that come in four types: *Leucoplasts* store proteins, oils, and starch; *amyloplasts* store only starch; *chromoplasts* store colored pigments; and *chloroplasts* carry out the process of photosynthesis. This crucial chemical action converts light energy into chemical energy and as such is responsible for supporting all life on earth.

Both plant and animal cells have a *Golgi complex*, although in plant cells, it is sometimes called a *dictysome*. Golgi complexes consist of many swollen tubules layered like a stack of pancakes. Their function is to package chemicals into balloonlike vesicles that eventually pinch away from the complex. Some will float along, remaining inside the cell; others will migrate to the cell's surface to eventually spill their contents to the outside.

Golgi complexes produce *lysosomes* and *microbodies*. Lysosomes are vessels that contain virulent acids that are eventually used to destroy the worn-out parts of a cell, like old mitochondria. The outer membranes of lysosomes must be continually renewed. Microbodies are also balloonlike organelles, but instead of containing harmful

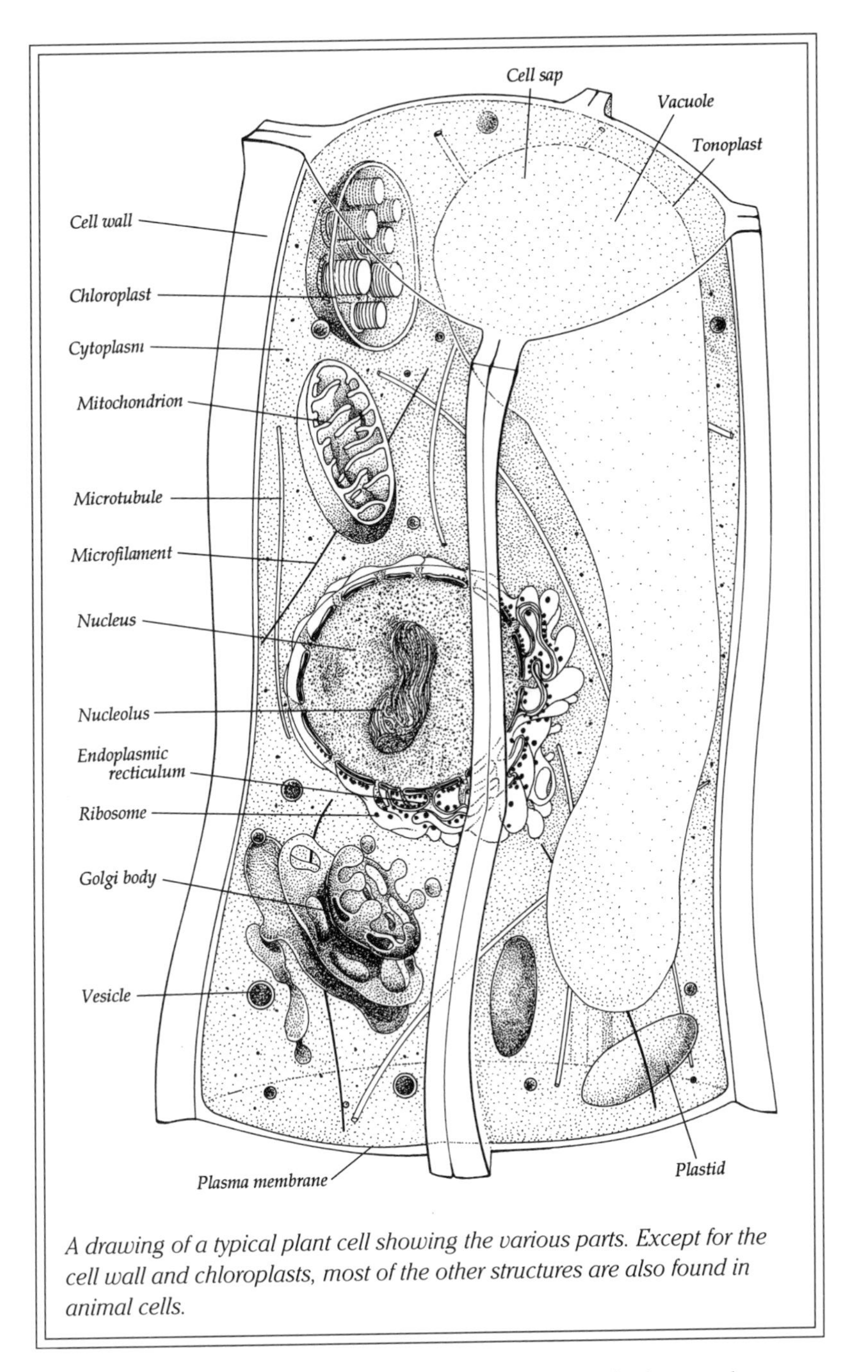

A drawing of a typical plant cell showing the various parts. Except for the cell wall and chloroplasts, most of the other structures are also found in animal cells.

chemicals, they contain enzymes that will eventually be used to turn the toxic chemicals left over from metabolism into harmless water and oxygen. Other microbodies are found in plant cells. Microbodies hold enzymes that convert fats and oils into sugars. These chemical changes are especially important during seed germination so that food stored in the embryo can be digested until photosynthesis is available to produce the seedling's food.

Both plant and animal cells have *vacuoles*, but in plant cells, vacuoles are large organelles that occupy up to 90 percent of a cell's volume. They are surrounded by a membrane called a *tonoplast*, which controls the direction liquids travel between the cytoplasm and the vacuole. Vacuoles are full of cell sap, a liquid that changes according to the age of the cell. Today, botanists believe that the vacuoles act as receptacles for metabolic wastes and other toxic chemicals used by the cell during its lifetime. Tonoplastic membranes keep these dangerous chemicals from returning to the living part of the cell.

Plant and animal cells contain *microfilaments* and *microtubules.* Microfilaments are thin protein fibers that act like a superstructure to help cells keep their shape. Microtubules are thin hollow tubes made from protein that are not only used to support the cell's contents, but also help in the movement of chromosomes during cell division.

The next time you eat a carrot or sow nasturtium seeds, remember that the cells that make up various parts of any plant are complicated units of life on a par with animal cells and, in the large scheme of things, go back to the beginnings of life on earth.

DIVISIONS IN THE PLANT KINGDOM

In recent years, the plant kingdom has been redefined. It once included bacteria, algae, slime molds (strange organisms composed of microscopic amoebalike cells that, under some conditions, actually move), funguses—represented by molds, yeasts, and mushrooms—and lichens, which are strange combinations of funguses and algae living together in symbiosis. Although sexual reproduction is found within this group, the sex organs and spore-producing structures are usually one-celled and very primitive, depending on water for many of their reproductive functions.

But in 1969, a new kingdom was created for the funguses and their kin, while the rest of the plants were shunted into two major divisions, the *bryophytes* (nonvascular plants) and the *tracheophytes* (vascular plants).

The first division, called bryophytes, number about twenty-four thousand species and include mosses, liverworts, and hornworts. These plants do not have true roots but absorb water directly into their outer cells, diffusing throughout the plant without a true system or network of vessels to carry the water. Their roots are called rhizoids, and their only function is to anchor the plant, not to absorb water or food. Essentially, these are land plants and most of them contain chlorophyll, giving them the distinct green color of plants. All bryophytes are small. The largest erect specimen grows to about twenty-four inches tall, and the smallest is nearly microscopic. The zygote, or fertilized egg, remains in the female sex organs for some time, during which it begins to divide and to form a mass of cells called a sporophyte. This mass eventually matures and releases spores that give rise to egg- and sperm-producing spermgametophytes that meet to form zygotes.

The second division, the tracheophytes (vascular plants), represents the first time that nature produced an independent sporophyte with true roots, stems, and leaves, signifying one of the most important steps in the evolution of the plant kingdom.

Today, tracheophytes number almost four hundred thousand species with more being discovered every day. Except for living fossils called whisk ferns (*Psilophyta*), the vascular plants have true roots, stems, and leaves, but they are further divided into two subgroups: the lower vascular plants that use spores for reproduction and the higher vascular plants that use seeds.

THE LOWER VASCULAR PLANTS

In addition to the whisk ferns, the lower vascular plants include club mosses, horsetails, and ferns. Club mosses are small herbaceous plants of the woods that are usually confused with true mosses. The most common genus in the United States is the club moss (*Lycopodium* spp.), often grown in shady gardens as a ground cover.

All the horsetails are in one genus, called *Equisetum*. They are usually green, have jointed hollow stems that resemble elongated poppit beads, and revel in damp soil. They were once used as scouring pads because of the high silica content in the stems. Horsetails can bring an unusual and primitive look to a garden but are widely considered pernicious weeds.

Two-thirds of today's living ferns are tropical, and all ferns are found growing in areas that evidence moist or wet soil. Because they reproduce by flagellated sperm cells that must swim to meet a female egg cell, ferns would quickly fade from the scene without water.

The life cycle of a typical fern involves spores, not seeds. The spores are found within spore cases, called *sori*, found on the undersides of fern fronds and often mistaken for insect infestations or disease by nongardeners.

When weather conditions are fairly dry and the spore cases are mature, their walls bend back and release millions of dustlike spores into the air. If it's raining, the sori stay closed, protecting the spores from destruction by heavy drops of water.

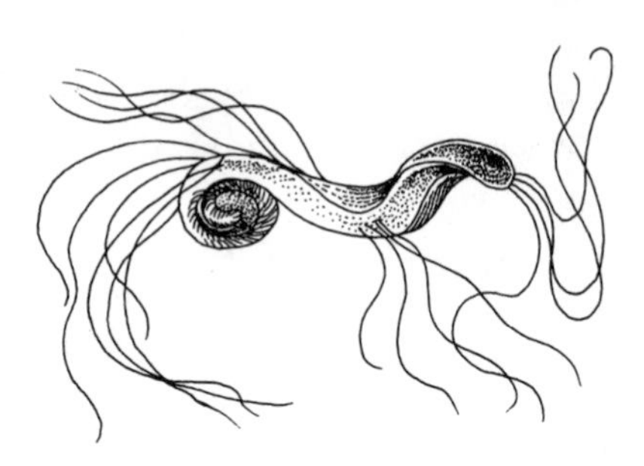

A fern sperm is made to swim through water.

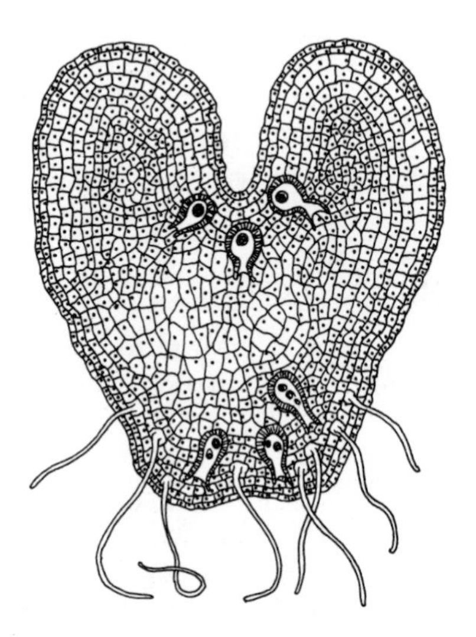

A fern prothallium with the male organs, or antheridia, at the bottom and the female organs, or archegonia, under the notch.

The spores drift to earth, sometimes traveling miles on a breeze. When the temperatures are reasonably warm and there is plenty of moisture present, a spore will develop into a flat, heart-shaped structure called a *prothallium,* about the size of an aspirin tablet.

Separate and distinct male and female organs soon appear on the lower side of a prothallium; one producing eggs, and the other, elegant sperms that swill through a thin film of water for fertilization. In a short time, an embryo fern begins to develop. For instructions on growing ferns from spores, see page 214.

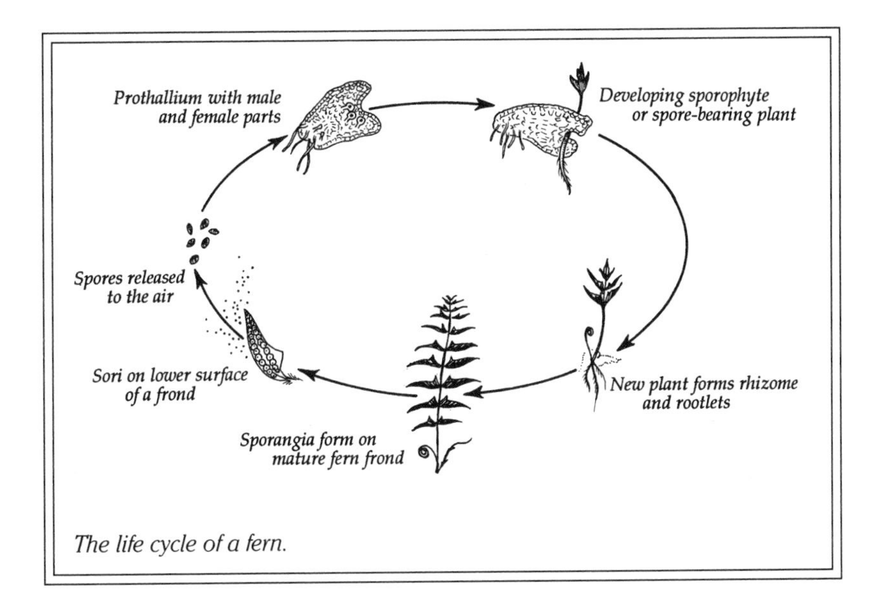

The life cycle of a fern.

PLANTS THAT PRODUCE SEEDS

Human life begins with the meeting of a sperm and an egg, each containing a set number of chromosomes that contain the hereditary plan for a new individual. It's the same with plants. The seed-producing plants, called *spermatocytes*, originated from lower forms of the plant kingdom through millions of years of evolution.

As far as the plant world goes, this evolution reached a high point when it was forced to respond to the changing environmental conditions that resulted from an aging earth. During the Devonian and Permian periods of the Paleozoic Era, most of the earth's soil was wet and swampy. The ferns and mosses were the supreme plants on the earth because their reproduction depended on the sperms swimming through water.

But when the continents began to drift apart and the land rose out of the lowlands, the earth's climate began to fluctuate, and the seasons were born. To adapt to these drier conditions, plants needed a new method of reproduction. The evolutionary outcome resulted in the earliest seed-bearing plants, called *gymnosperms* from *gymno*, or naked, and *sperm*, or seed, a word that combines to represent plants that bear naked seeds. Gymnosperms are represented by two groups: the *gymnosperms* or pines, firs, cycads, and so forth; and the *angiosperms* (*angio* meaning "covered"), or flowering plants.

THE GYMNOSPERMS

Gymnosperms were able to latch on to the evolving earth because they grew extensive root systems that hold plants in place while extracting water and minerals from the ground. Gymnosperms also developed complex transportation tubes called *xylem* and *phloem* to carry liquids throughout the plants. And best of all, they developed a process of fertilization where the sperms become dustlike pollen, not needing water for fertilization. Finally, these plants produced the seed, a plant embryo within a protective cover, so germination could be delayed until environmental conditions became favorable.

The life cycle of a gymnosperm is easily represented by looking at the white pine (*Pinus strobus*). Every spring, pollen-bearing structures, or male cones, appear at the ends of most lower branches of a mature white pine. These cones cluster just below the current season's crop of new needles. When the male cones are mature, they release clouds of yellow dust made up of tiny-winged grains of pollen. These grains float on the air currents to the treetops, where the female cones are found. They also find their way into countless open windows of nearby houses and carpet everything with a layer of yellow dust.

When individual eggs are fertilized by pollen, they begin the lengthy process of producing seed. Seed generation takes thirteen months and then another twelve months or so for the pine embryo and the seed to mature. When that happens, the cones point to the ground and open, releasing winged seeds that gently glide to the earth. These seeds remain unchanged for the winter and then germinate under the warm sun of the following spring, giving life to a new baby pine. Compared with the previous plant life cycles, it's fairly sophisticated, but effective enough to populate the earth.

THE ANTHOPHYTES

The angiosperms, or flowering plants, belong to a division called the *Anthophyta* (a word that means "enclosed seeds"). They appeared on earth about one hundred million years ago, and today nearly 75 percent of all the plant species on the earth are flowering plants.

THE FLOWER

Although seeds rate very high on the scale of the incredible, flowers hit the top. Whether viewed by color, form, fragrance, or the sheer combination of all these traits, flowers possess a beauty that has charmed every society ever to tread this earth.

Some flowers are bizarre. Even a writer on the level of H. P. Lovecraft or Edgar Allen Poe could never have conceived a flower as strange as the *Rafflesia.* This parasitic plant from Malaya and the East Indies has vegetative organs consisting of thin threads that penetrate the tissues of the host plant. At this stage, the plants are so tenuous that only when the flower appears is *Rafflesia* really visible. And what a flower! Currently known as the largest blossoms on earth, their fleshy girth measures more than three feet in diameter. Four or five conspicuous petals are splotched with cream-colored spots on a red background and, eventually, they produce a large berry.

Another rarity, the flower of the orchid *Ophrys speculum,* actually takes advantage of the powerful sex drive found in certain male wasps for pollination. Because the flower closely resembles and smells like a female wasp, the male actually copulates with the flower and, in so doing, throws pollen around, resulting in the fertilization of the orchid.

A receptacle of a typical flower is held on a shortened stem called a *pedicel.* Before opening, it is protected by an outermost ring or *whorl* of modified leaves called *sepals.* The second whorl is made of *petals,* collectively called the *corolla* of the flower.

Petals are also modified leaves. Some are brightly colored to attract insects; others act as pollinators. Some flowers, like poinsettias, completely lack petals. The leaves that surround the tiny poinsettia blossoms become highly colored bracts to attract pollinators. Some petals actually have lines that act like arrows that guide insects directly to the center of the flower. Sometimes, flowers like the Lenten rose (*Helleborus orientalis*) have modified the petals into an entirely different structure, called a nectarie, which acts like a gland that secretes sweet nectars. To remedy this defect, the sepals act as petals.

Inside the whorl of petals there is a whorl of *stamens,* modified leaves that form the male reproductive part of the flower. Most stamens have a long stalk or *filament,* with a swollen tip known as the *anther.* Pollen grains are produced inside the anthers in *pollen sacs.* When these are mature, they open to release the pollen.

The whorl at the center of the flower is called the *carpel.* These are modified leaves that make up the female reproductive part of the plant. Collectively, carpels are called a *pistil.*

At the top of the pistil is a sticky area called the *stigma,* the place where pollen collects. The stigma is connected to the *ovary* by a *style.* Basically, when pollen hits the stigma, a *pollen tube* is produced and leads down to the *ovary* carrying the *sperm* to meet the *egg,* thus producing the seed. At the same time that this process occurs, the ovary, or other floral parts, mature to form a fruit that surrounds and protects the seeds until they are ready for dispersal.

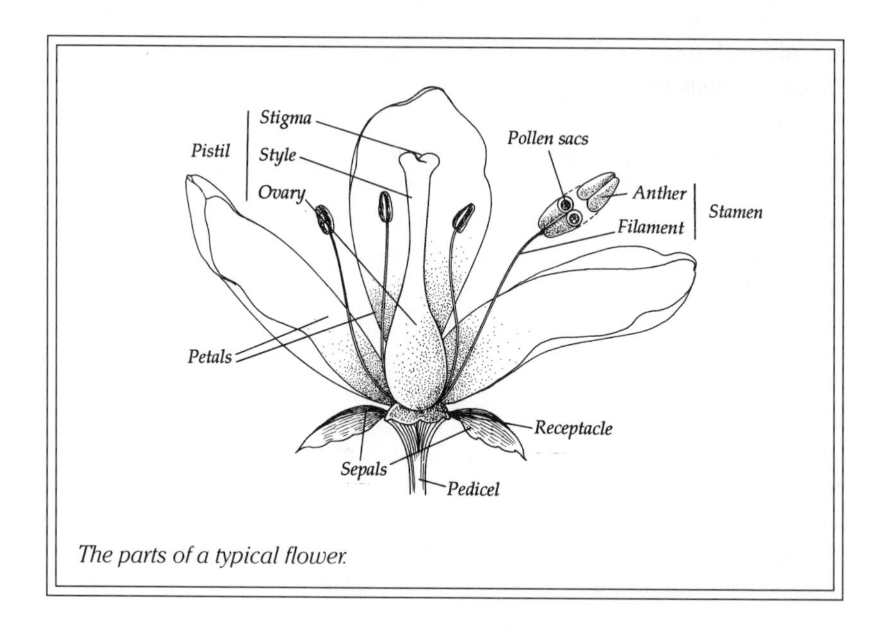

The parts of a typical flower.

Not all flowers are the same. Just as there are minor variations found in any group of people, there are variations found in flowers. Because the construction of the whorl is not constant, flowers are differentiated by the number of parts found in the whorl. And obviously, there are almost an infinite number of color variations in petals. There is no constant symmetry in flowers: Some, like many orchids, have an obvious top and bottom; others, like a daisy, are round. In some flowers, the various parts are fused together; in others, these parts are separate. There are also differences in the shape of the ovaries. And flowers can appear as individuals or are found in clusters or groups called *inflorescences.*

Finally, flowers that have all four whorls are known as *complete,* whereas flowers that lack one or more whorls are logically called *incomplete* flowers.

When it comes to sexual identification, all complete flowers are known as *perfect* flowers because they contain both male and female parts. *Imperfect* flowers are either male or female; one has stamens, the other has carpels, but never both at the same time. The rose has both male and female flowers on the same stem, so it's known as a *monoecious* plant (from *mono,* meaning "one," and *oecious,* meaning "house"). Other plants are called *dioecious* (*di,* meaning "two"), and the imperfect male and female parts are found on separate plants. Holly trees are dioecious, and you need both a male and female plant in a garden to ensure pollination.

Mitosis is the usual method of cell division in which the mother cell divides into two daughter cells, but all wind up with the same number of chromosomes. Cells use mitosis to repair old cells or produce new cells for healthy growth.

MONOCOTS AND DICOTS

Botanists divide plants into two groups: *monocotyledons* or monocots (*mono*, meaning "one") and *dicotyledons*, or dicots (*di*, meaning "two"). There are primary differences between the two groups, although only the truly observant gardener ever notices.

Monocots and dicots share the same basic plant, consisting of roots and stems, leaves, flowers, and fruits. But in the dicot's stem, for example, the vascular bundles that deliver water and nutrients to the leaves are arranged in rings, while in monocots they are scattered throughout the stem. As a result of this arrangement, monocots do not develop woody tissues and have no secondary growth characteristics. Except for palm trees (not really trees and having a completely different type of growth), if monocots become too large, they simply fall over.

Monocots have fibrous root systems, like grasses, and dicots have one main root called a tap root, as evidenced by the carrot.

It's easy to spot the difference by the leaf structure. Dicots, like a maple tree, have many veins that form a netlike design across the entire leaf. Monocots have parallel veins, like those in a hosta leaf.

When it comes to flowers, monocots are grouped primarily in threes; each displays three petals and three stamens, whereas the flowers of dicots are arranged in groups of four or five. Trilliums and lilies are monocots, but most of the familiar seed plants are dicots. Last, a monocot has one seed leaf, or *cotyledon*, and a dicot has two.

SEXUAL REPRODUCTION OF FLOWERING PLANTS

The seed-bearing plants of division four are the culmination of plant evolution. In all the lower plants, the fertilized egg develops immediately into the mature sporophyte, whereas the zygote grows for a time in seed plants but goes into a dormant condition to form the seed.

Before the emergence of seeds, the production of the gametophyte, or the sexual generation of a plant, would alternate with the sporophyte, or asexual phase. The process works reasonably well, but it means that the dispersal of the plant is very limited, and there is little chance of spontaneous genetic changes. Remember that Mendel did his genetic discoveries working with peas, not ferns or mosses.

In the development of seed plants, pollen mother cells divide by a process called *meiosis*, resulting in the pollen's being *haploid*, or having half the number of chromosomes characteristic of the species. The same thing happens to the embryo sacs that contain the eggs. So

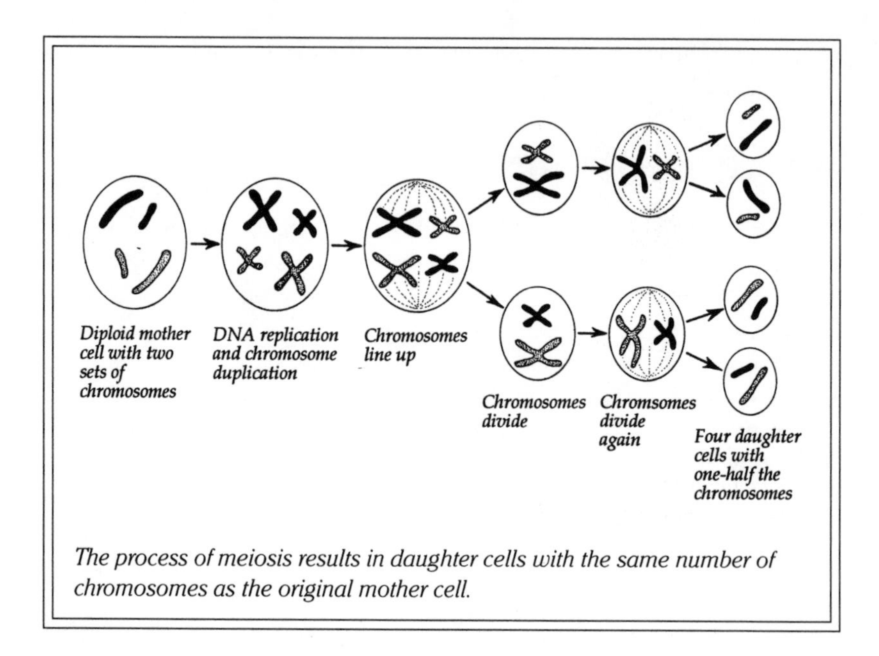

The process of meiosis results in daughter cells with the same number of chromosomes as the original mother cell.

when the sperm and egg fuse at fertilization, the act restores the double, or *diploid*, chromosome number in the new embryo or seed.

FORMATION OF MALE GAMETES

Pollen grains form in the pollen sacs of the flower's anther. There, a *microspore mother cell*, each with a full complement of chromosomes found in that particular plant, begins to divide by meiosis into four haploid microspores, each with half the number of chromosomes. Using mitosis, they develop into four pollen grains, a two-cell male gametophyte, consisting of a generative cell and a *tube cell*. When the pollen lands on a stigma, the generative cell divides by mitosis into two sperms, and the tube cell develops a tunnel that carries the sperms to the egg cell within the ovary.

Individual pollen cells are a thing of beauty. Each species has unique qualities that identify them as easily as fingerprints identify human beings. Some have tiny spines to help them cling to pollinators; others have tiny wings to help them fly through the air.

FORMATION OF FEMALE GAMETES

Gametes are the sex cells. Female gametes are formed in the *ovules*, oval masses of cells that are attached by a thin stalk to the inside cavity of the ovary. Depending on the genus, some ovaries, like the orchids, contain millions of ovules, whereas peas, for example, contain only a

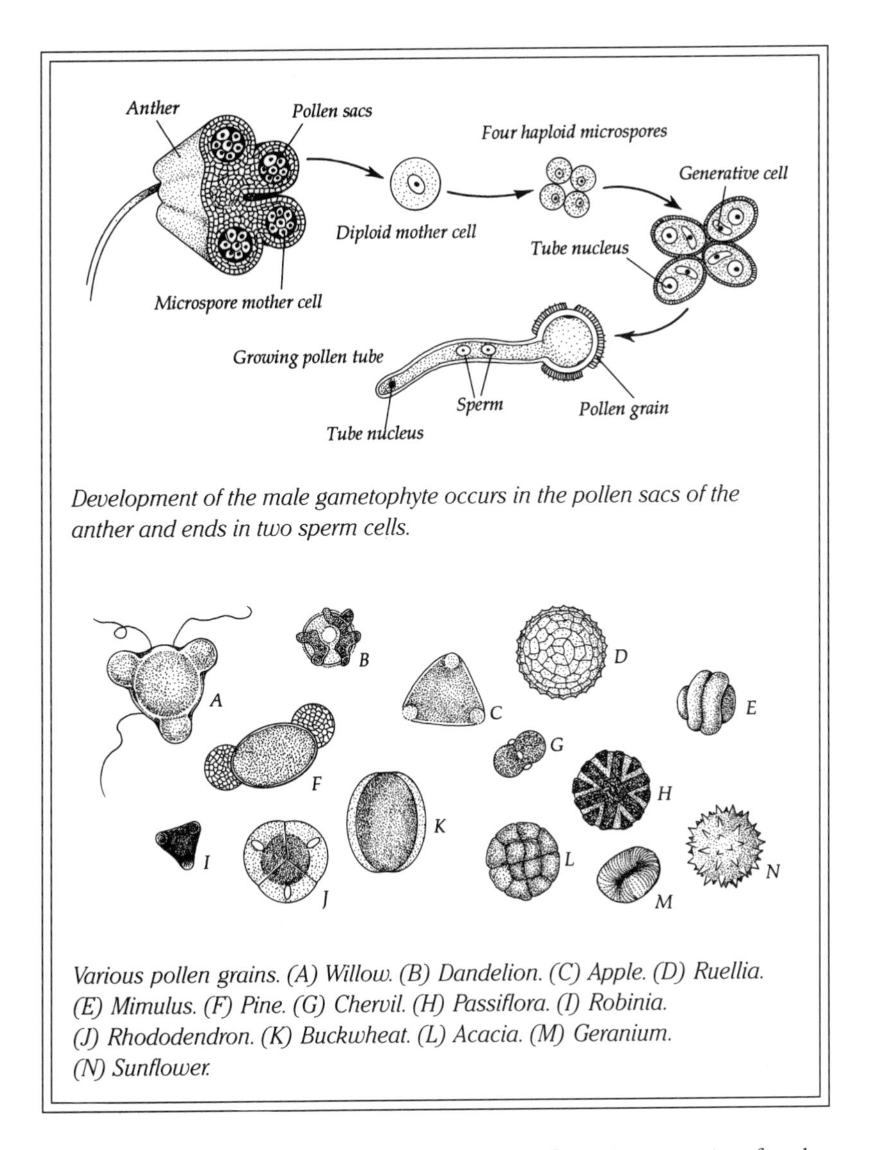

Development of the male gametophyte occurs in the pollen sacs of the anther and ends in two sperm cells.

Various pollen grains. (A) Willow. (B) Dandelion. (C) Apple. (D) Ruellia. (E) Mimulus. (F) Pine. (G) Chervil. (H) Passiflora. (I) Robinia. (J) Rhododendron. (K) Buckwheat. (L) Acacia. (M) Geranium. (N) Sunflower.

few. At the bottom of the ovule is a *micropyle*, a tiny opening for the male pollen tube to enter. The process begins with a megaspore mother cell that contains the normal number of chromosomes for that particular plant.

Using the process of meiosis, the megaspore mother cell divides into four haploid megaspores. Three of the four degenerate, leaving one. The remaining megaspore divides three times using mitosis, resulting in a female gametophyte with eight haploid nuclei. The nuclei then migrate to different parts of the megaspore: Three go to the top, called antipodal cells; two stay in the center, called polar nuclei; and three go to the bottom and become the egg apparatus. The entire megaspore is now called the embryo sac.

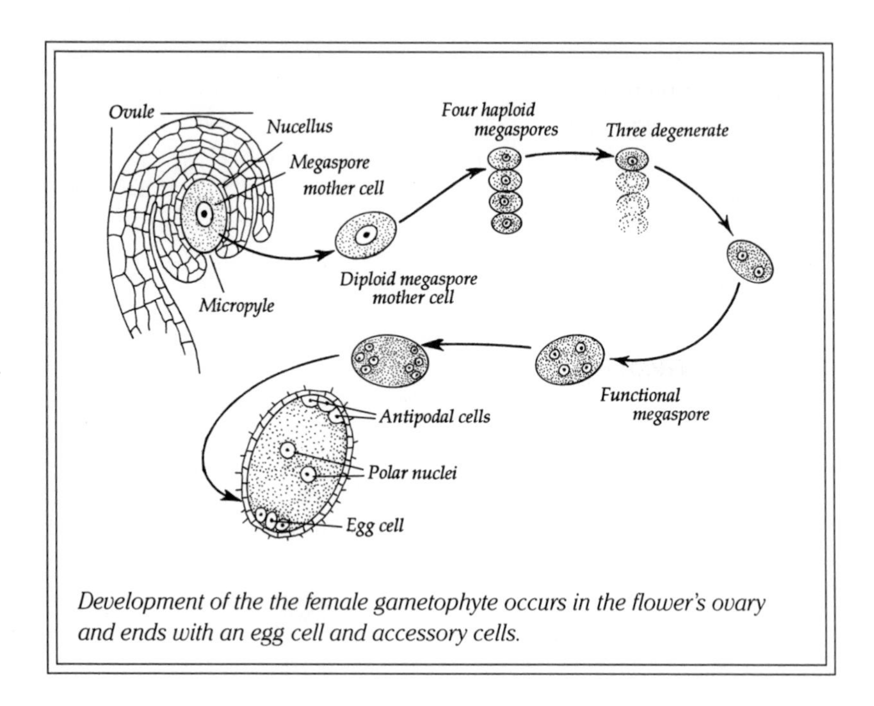

Development of the the female gametophyte occurs in the flower's ovary and ends with an egg cell and accessory cells.

THE FERTILIZATION OF A FLOWER

Remember the plight of the earlier plants on earth? They needed water to aid in fertilization. Water is still used, of course, for a few of the anthophyta that dwell in water and depend on their environment to aid in pollination. Eelgrass (*Vallisneria*), a member of the grass family, is a marine herb that is always submerged. The flowers are perfect, with the sexes alternating in two rows on one side of a flattened stem. The male flowers are reduced to one anther and produce threadlike pollen that reaches an ovary by floating through the water. The seeds have no endosperms and are dispersed by the movement of water and fish.

There is also a monoecious South African water lily (*Nymphaeaceae*) that uses water to great effect for pollination. Some of the flowers open in an initial female stage without pollen and some in a second, or male stage, in which vast amounts of glorious yellow, nutritious pollen are spread before flying insects. Confused by all the food, some insects will fly from a flower in the male stage to a flower in the female stage, never noticing the lack of pollen at the second flower. The immature stamens circle a pool of water in the flower's base, and while looking for pollen, the searchers slip to a watery grave, helped along by wetting agents in the water that make sure every visitor is a victim. The flowers close at night, and the pollen washes off the corpses, then settles to the bottom of the pool where the stigma is found. The next morning the pollinated flower becomes a male and produces pollen to help another flower with fertilization.

Most flowers depend on the wind for distributing pollen. Others depend on insects to move pollen, including honeybees, bumblebees, wasps, flies, beetles, gnats, and ants, not to mention moths and butterflies. Still others are pollinated by bats and hummingbirds or even by some small mammals like mice.

THE DEVELOPMENT OF THE EMBRYO

Once the sperm reaches the egg, two things happen. The first sperm travels down the pollen tube and actually fuses with the egg to form a diploid zygote, or fertilized egg. The second sperm goes to cell center and fuses with the two polar nuclei left over from the female divisions, forming a triploid primary endosperm cell.

Immediately upon fertilization, chemical messengers go out to the flower, and several changes begin to happen. First, the petals begin to die, and the flower is no longer receptive to new pollen. Then, the zygote begins to develop into the plant embryo while the primary endosperm cell works to produce endosperm tissue for support of the embryo during germination.

Within the developing seed coat, things really begin to move along. While the two cells of the egg apparatus and the three antipodal cells degenerate, the endosperm tissue continues to develop while the zygote divides into two cells, one smaller than the other.

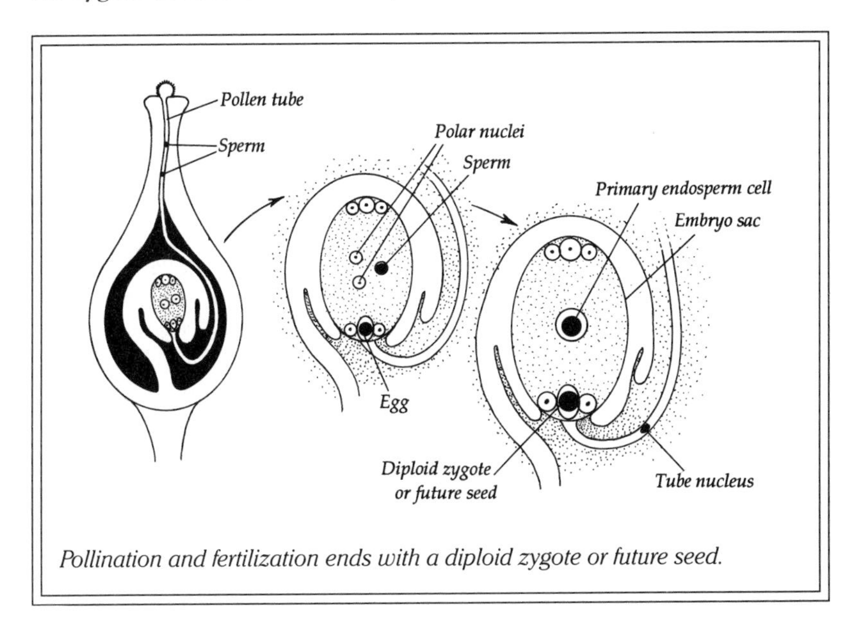

Pollination and fertilization ends with a diploid zygote or future seed.

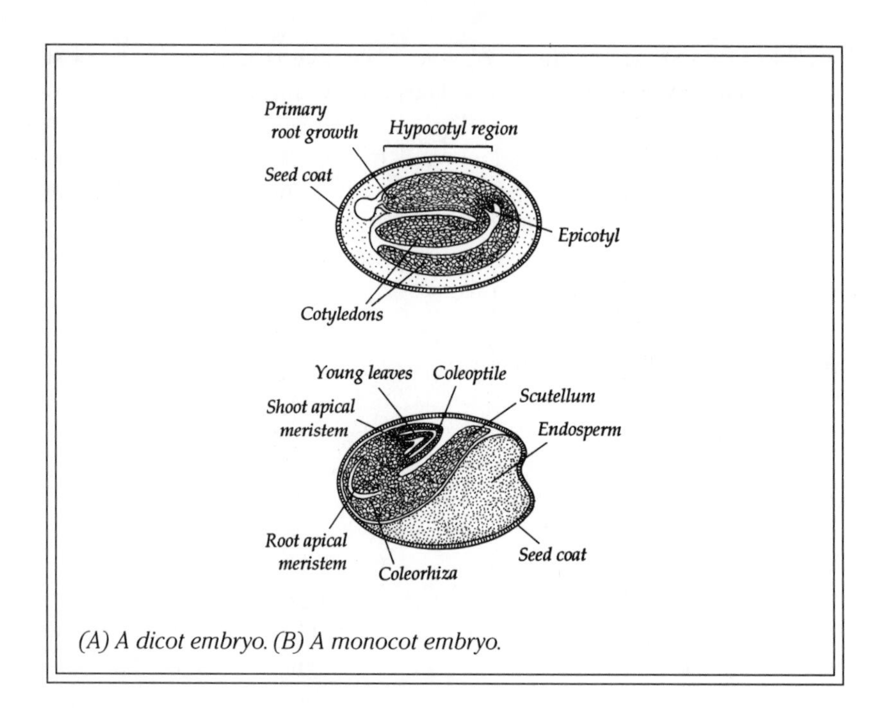

(A) A dicot embryo. (B) A monocot embryo.

The smaller zygote goes through numerous mitotic divisions. If it's a monocot, then one seed leaf, or cotyledon, develops. If it's a dicot, then two seed leaves develop. These seed leaves contain food reserves. (That's why it isn't necessary to feed seedlings grown in nourishment-lacking mediums until the true leaves appear.) The larger zygote cell becomes the suspensor, a column of cells that conduct nutrients from the endosperm, or food bank, to the developing embryo.

But there are more complications: In addition to the two cotyledons in dicots, the mature dicot embryo will have an epicotyl with a plumule at its tip. The epicotyl eventually becomes the shoot, and the plumule becomes the leaves. Below the cotyledons is the hypocotyl, and at the top of the hypocotyl is the radicle, the part of the embryo that eventually becomes the root system.

Monocots have only one center of division, and it develops at the top of the growing monocot embryo, producing only one cotyledon, known as a *scutellum.* The *coleoptile* is a special structure that surrounds the shoot tip and protects the young stem and leaves as they emerge from the soil. The *coleorhiza* is like a knife sheath that protects the root tip as it penetrates the soil.

THE DEVELOPMENT OF THE SEED AND FRUIT

While all the activity is going on within the ovule, a seed coat is forming to protect the contents from danger. Depending on the species, seeds are hard or soft, and many have various projections that help the seed coat to absorb water or to aid in the attachment of the seed to an animal's fur or a person's clothing, thus helping in dispersal (see page 61).

Remember the micropyle—that small opening that allowed the pollen tube to enter the embryo sac? It's still found in almost all seeds and becomes a tiny opening in the seed coat, where it serves another function in germination (see page 44). Next to the micropyle is the *hilum*, a scar on the seed that marks the point where the ovary's stalk was attached to the ovule.

During all this activity, the ovary is turning into the fruit. Whether it's red like an apple, green like a tomato, or grains of wheat, all the seed containers are fruits because they developed from ovaries.

Fruits come in three categories: *Simple* fruits, like the apple, develop from one pistil; *aggregate* fruits, like the strawberry, develop from many pistils in one flower; and *multiple* fruits are the result of the pistils in separate flowers, the best example being the pineapple.

Just to make things more difficult, there are fruits without seeds, developed in a process called *parthenocarpy*. In this case, the fruit contains no embryo and, hence, no seeds. Seedless cucumbers, bananas, seedless grapes, and seedless watermelons spring to mind when discussing parthenocarpy. These fruits are the result of plant breeding and have nothing to do with nature but a great deal to do with the attractiveness of food. In some cases, artificial plant hormones are sprayed on the ovary to force maturity without actual fertilization.

With seed plants, sexual reproduction reached a high point of efficiency. Nobody can doubt their success in becoming the dominant form of the earth's vegetation. Seed plants have not only spread over great land masses, but also, up to now, survived through long periods of poor environmental conditions. As for the future, that's anybody's guess.

A tulip blossom (Tulipa *spp.)*

CHAPTER 2

Pollination Leads to Seeds

Nature tends to make the wonders of the vegetable world look temptingly easy to come by, especially when you forget the millions of years that went into perfecting only the slight changes in floral structure. But add to the mix the various adaptations found in living pollinators, and you'll mutter the words of the King of Siam, "It's a puzzlement."

For seeds to form, the flowers must be pollinated. That can happen in several ways, many of which are extraordinarily clever, and all show a high degree of sophistication.

For years, gardeners, philosophers, and farmers have known about the relationship between pollen and the formation of fruit. Early on, Egyptian reliefs showed the dusting of fruit-bearing date palm flowers with pollen, but centuries passed before the link between the two was finally established by Rudolph Jacob Camerarius (1665–1721). In 1694, Camerarius, a German botanist and physician, demonstrated that unless pollen touched a stigma, fruit would not develop. In his work he described the stamen as the male organ and the ovary as the female organ, emphasizing their relationship in the formation of seeds.

Today, after centuries of observation, we've learned a great deal about pollination and know that it's carried out by five pollinators, the most important being the wind (or *anemophily*). Next comes pollination by insects (*entomophily*) and birds (*ornithophily*), followed by the relatively uncommon pollination by bats (*chiropterophily*), and finally, pollination by water (*hydrophily*).

POLLINATION BY WIND

In the spring, if you live anywhere near blooming oak trees, yews, or gymnosperms like pines, you know about the pollen that blows in through windows, soon giving everything a coat of yellow dust. It's really impossible to count the number of pollen grains produced by one tree. So you can imagine the effectiveness of wind pollination—like hitting a pin with a hammer. From nature's viewpoint, however, pollen is cheap to produce.

The earliest seed plants on the earth were probably pollinated by the wind, and today's most modern gymnosperms still are. Today, botanists think that the original flowering plants or angiosperms were pollinated by insects, but over the course of evolution, many of these plants have turned to the wind.

It's easy to recognize a wind-pollinated plant. They have no nectar, flower fragrance, or brilliant colors to attract animal pollinators. Instead, their floral structure is perfectly suited to the wind. Grasses are wind-pollinated, and although floral structures are still visible, they are tiny and serve no purpose. The anthers are often suspended from long filaments, hanging open to the wind and pollen. The stigmas are often feathery or branched in appearance, and all are open to the air and never protected by petals. Finally, pollen grains are usually small and very smooth and are produced in copious quantity.

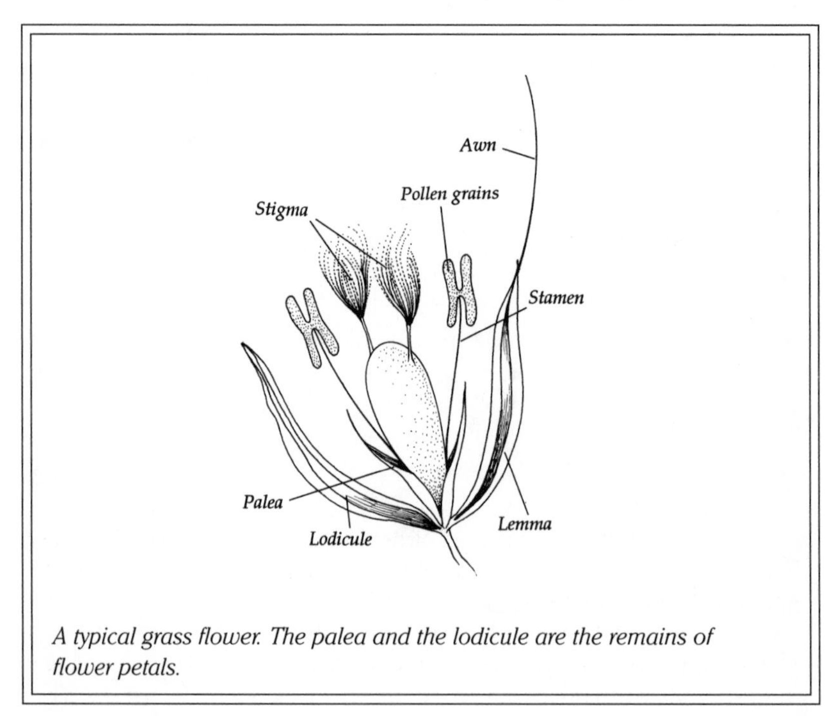

A typical grass flower. The palea and the lodicule are the remains of flower petals.

Nectar guides in flowers. (A) Nasturtium. (B) Marsh marigold.

POLLINATION BY INSECTS

Flowers that use insects as pollinators lure them for food, with either nectar or sometimes pollen. Nectar is mostly sweetened water, with a 25 percent glucose level.

Usually, insects are attracted to flowers by color, scent, or sometimes both. Although many mammals see only in black and white, bees, for example, can see yellow, blue, and purple. And bees associate these colors with nectar.

Flowers that attract insects are usually held erect with a pronounced rim for visitors to land upon. The petals are often marked with dots or lines, called nectar guides, that lead to the flower's center. People and bees see different wavelengths of light, and the flower of a marsh marigold appears pure yellow to you or me. But to a bee, the petals have distinct lines that converge at the flower's center.

Insect-pollinated flowers are divided further into four categories: pollen flowers, bee flowers, hovering insects, and fly flowers.

Pollen Flowers

Pollen flowers do not have nectar but produce a lot of pollen food, a primary food source not only for bees and beetles, but also for other insects. Take a flashlight out to a lily flower at night, and you'll be amazed at how many insects gather there to eat. Pollen flowers include poppies (*Papaver*), roses (*Rosa*), and lilies (*Lilium*).

Bee Flowers

Bee flowers (called *hymenoptera* flowers by botanists) have special adaptations, so only bees and beelike visitors find it practical to search for honey.

With some flowers, snapdragons (*Antirrhinum*) or monkshood (*Aconitum*), for example, the corolla is so firm that only a heavy insect like a bee or a bumblebee has the weight to open it. Red clover (*Trifolium*) imbeds the nectar so deeply within the blossoms that only bumblebees have a tongue long enough to reach the prize.

Lepidoptera Flowers

Most flowers in this class are visited by insects that hover in the air while they eat. The nectar is usually hidden in spurs; the classic example is an orchid from Madagascar, called the Christmas star (*Anagraecum sesquipedale*), a flower with a foot-long spur. When it was discovered, Darwin predicted that an insect existed with a foot-long tongue. In 1828, botanists found the moth that pollinates this orchid.

Because butterflies lack a highly developed sense of smell, butterfly flowers are often brightly colored, including many with petals of pure red or orange. Their odor is generally light, pleasant, and very agreeable, like the smell of a meadow basking under a summer sun.

Moth flowers are usually white or, at best, have only faint tints and are often drab and insignificant. Their perfume, however, is strong and heavy, often having a very sweet and soapy smell. Except for orchids, most nocturnal flowers are tubelike, with the nectar hidden deep within the blossoms. And they often close during the heat of the day.

Fly Flowers

Except for a few exceptions, fly flowers (or *dipteras*) attract insects mainly by smell—usually that of rotting carrion—or sometimes in combination with the smell have the color of rotting flesh. A major garden exception is the speedwell (*Veronica*), a flower that is designed to be pollinated by hover flies, and it's amazing to find that's exactly what shows up at these blossoms.

A typical fly flower, here the toad cactus (Stapelia variegata).

Favorite fly flowers are the starflowers or toadflowers (*Stapelia*) from South Africa. The fly pollinators are attracted by both hairy and dark-colored petals, often looking like flesh and smelling like dirty sneakers fried in rancid butter.

POLLINATION BY BIRDS

Bird flowers attract hummingbirds, honeysuckers, and sunbirds. Flower colors are more than 80 percent red and usually very bright. They lack scent but have an abundance of nectar. About two thousand species of birds regularly visit flowers, and about two-thirds of the species depend on flowers for most of their food.

POLLINATION BY BATS

Bat flowers usually open at dusk. Some smell of overripe fruit; others give off the odor of mice or rats. The flowers are positioned to make it easy for the bats to visit, and they are often suspended on long stalks. An exception to this scheme is the saguaro cactus (*Carnegiea gigantea*). Here the smell is not unattractive, and the Sanborn long-nosed bat

(*Leptonycteris sanborni*) approaches the flower from above, then uses its wings like parachutes, hovering over the blossom for a quick sip before flying away.

POLLINATION BY WATER

Water pollination is rare but happens on occasion. One example is the popular aquarium plant eelgrass (*Vallisneria*). The plants are dioecious with male and female flowers on separate plants. As buds, the female flowers are well below the water's surface. As they mature, the flower stem (or pedicel) spirals up, eventually reaching the surface, where the flower floats within a dimple on the water's film. Meanwhile, tiny male flowers, consisting of two stamens attached to a boatlike float, are released by the hundreds and rise to the surface. They drift about until surface tension draws them into a dimple inhabited by a female flower. When the anthers touch the stigmas, pollination occurs.

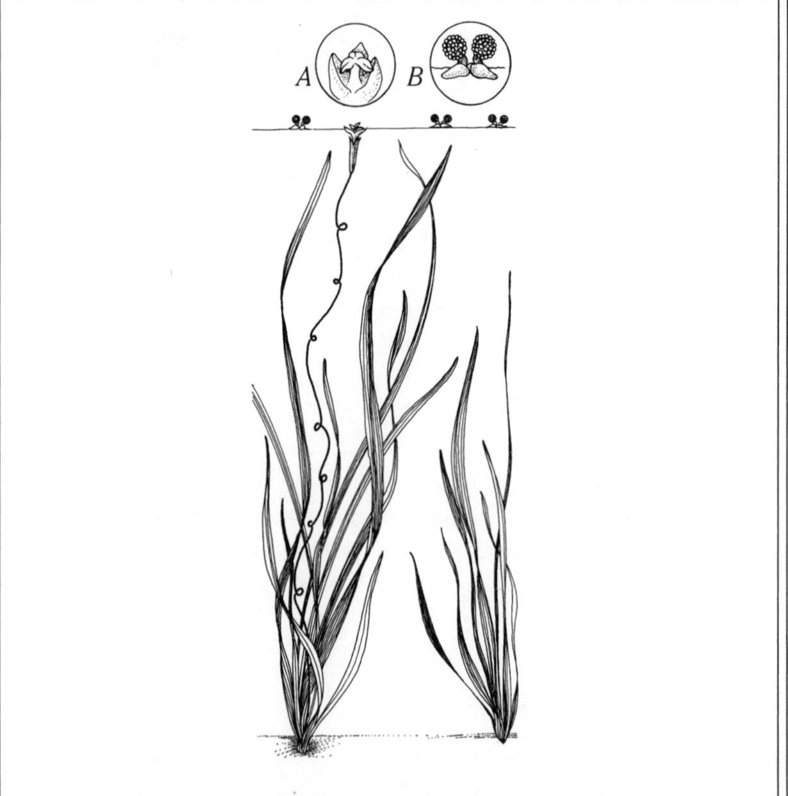

Eelgrass (Vallisneria americana) *showing the female flower (A) and the free-floating male flower (B).*

AN EXAMPLE OF A SPECIAL CASE

The Yucca

A great number of American plants were imported to Europe during the sixteenth century, including tobacco, potatoes, nasturtiums, sunflowers, and yuccas.

In 1629, John Parkinson (1567–1650), one of England's great botanists and the apothecary to King James I, published his monumental book *A Garden of Pleasant Flowers.* He wrote of the Indian yucca as "a rare Indian plant [that] hath a great thicke tuberous roote (spreading in time into many tuberous heads) from the head whereof shooteth forth many long, hard, and guttered leaves, very sharpe pointed of a grayish greene colour, which doe not fall away, but abide ever green on the plant." Along with this accurate description is a beautiful woodcut showing a branch of flowers engraved by a hand that is very fine indeed.

Following the description is the news that the yucca first came to England from the West Indies by way of a servant who gave it to his master, Thomas Edwards, an apothecary of Exeter, who kept it until his death.

A yucca moth (Tegeticula yuccasella) *hovers above the blossoms of an Adam's needle* (Yucca filamentosa).

The word *yucca* is said to come from *yuca*, Spanish for the manifhot (*Manihot esculenta*), a major source of the bitter cassava, itself based on a Taino Indian word for this plant. Although used in error, the name stuck. There are about forty species native to the warmer regions of North America, many of which have now spread throughout the country. They all belong to the Agavaceae family.

Most people are familar with these very imposing plants with their sword-shaped leaves—often possessing very sharp points—because the yucca has been used in the landscape for hundreds of years. But few people realize that its tall spires of white, summer-blooming, bell-shaped flowers are pollinated in its native desert home at night by a species of moth. The same agent of pollination—though sometimes of a different species—does the job in many of the new areas where the yucca has settled.

Yuccas are not specifically night-blooming plants because the flowers are open during the daytime hours. But the flowers exhibit nyctinasty, the art of being nyctitropic, or in other words, they move about at night.

During the day, the white, six-petaled blossoms hang down like bells at rest. At dusk, they turn up to the evening sky, open wide, and release a sweet soapy smell to the night air. The reason for this behavior comes in the form of a particular genus of moths, the yucca moths, *Tegeticula yuccasella* (synonym *Pronuba yuccasella*).

According to Holland's *The Moth Book*, in 1872 Professor C. V. Riley discovered the relationship between these moths and the yuccas. He proved that the pollination of the flowers and the development of fruit was not an accidental meeting of insect and pollen. Instead, he proved it was the result of an insect purposely collecting pollen with a mouth modified by nature for the task and the subsequent application of the pollen to the stigma of a yucca flower. All this was done for the yucca to produce seed and the moth progeny to have food upon hatching.

The female moths enter the yucca flowers at night and with a specially modified first joint of their extended jaw or maxillary palp are able to roll the pollen in a tight ball. This is held under the moth's head as it flies to another flower. Here, the moth clings to a few stamenal filaments and with her ovipositor, neatly lays her four or five eggs in the side of the pistil. She then pushes the pollen ball into the funnel end of the stigma and ensures pollination.

A few days later, the moth eggs hatch, and larvae move into the blossom's ovary, where they consume eighteen to twenty seeds and gain enough strength to eat through the ovary's wall to the outside. Once out, they use a self-produced thread to lower themselves to the ground, where they burrow into the earth to complete a juvenile existence and emerge as full-grown moths. The hundred or so seeds not consumed by the larvae continue to ripen enough to guarantee the continued existence of the yucca. Because most yucca flowers are incapable of self-pollination—their pollen is puttylike, and the anthers and stigma would never touch—without the intervention of this particular moth, the plants would have perished eons ago.

Tegeticula yuccasella pollinates the blossoms of *Yucca filamentosa* in the East and does the same in the West for *Y. angustifolia*, whereas *Y. brevifolia* (the Joshua tree) is pollinated by *T. synthetica,* and *Y. whipplei* is polinated by *T. maculata.*

Seeds and blossoms of the spider plant (Cleome hasslerana).

CHAPTER 3

In Search of Heredity

Johann Mendel was born an Austrian, in 1822. At that time, most of the world was nonindustrial. Securing food took priority over other forms of economic activity because wants were related directly to survival. Agriculture was the primary mode of existence, and the industry employed and supported most of the world's population.

When Mendel, the son of a farmer, was eight years old, the British agricultural system was the most efficient in the world. By 1843, when he entered the Augustinian monastery at Brno, as the novice Gregor, industrialization was spreading, and the relative importance of agriculture was beginning to decline. John Deere introduced the steel plow in 1837, and that same year, Hiram and John Pitts of Maine introduced the first successful thresher. In the 1840s, chemical fertilizers began selling in France and Germany. And in 1848, gold was discovered in California.

In 1853, the young monk left his monastic cell to teach natural history at a local secondary school as an assistant teacher because he never passed the higher examinations necessary for the job as required in Vienna. He was elected abbot of the monastery in 1868, a position that caused him endless worries and led to one of the great intellectual tragedies of the Western world. For eight years before assuming this position, he carried on experiments in breeding peas, hawkweeds, and bees. And because he was a shy man and never made great noises about his discoveries, the world never heard of his experiments.

In 1865, he delivered his findings in a paper, "Experiments with Plant Hybrids," that he read at a local botanical society. In 1866, his papers were published, and the world rolled on.

Even though Mendel's duties at the monastery continued to take up much of his time, he went on with his experiments in the garden, anticipating Oscar Hertwig's discovery that fertilization of an egg involved only one male sex cell, but never publishing his results.

Sixteen years after Mendel's death, his discoveries became known by science. Three scientists, the Dutch botanist Hugo De Vries, K. E. Correns, and Tschermak-Seysenegg, were all involved with genetic work at the same time, and in 1900, they all discovered Mendel's work almost at the same time. Because they were all honest men, as a result of their researches, the long-forgotten Mendelian experiments were introduced to the scientific world.

"A few bags of seeds were his tools," wrote Mendel's biographer Ingo Krumbiegel. His goal was to cross two plants of the same genus, differing from each other in a few obvious traits, like flower color, the texture of a pea, or the size of a particular plant.

Using the garden pea, he chose the best examples for his experiments, tested their purity for two years, and only then began to cross-fertilize the flowers. Once the flowers were pollinated, he covered them in paper bags, preventing any wayward insects from compromising his work. By crossing two pea plants that differed in only one major characteristic, he discovered the principle of dominant-recessive heredity, in essence, that some genes are dominant, like yellow seeds versus green seeds. These results immediately flew in the face of the prevailing belief of blending.

For all the time that Mendel was out in the monastery garden, the majority of the world held the following belief: Parents produced hereditary fluids that mix together to form children exhibiting blended characteristics. The term *blood relative* is left over from this view of heredity. Such blending would always produce offspring that were homogenized versions of their parents, children that stood halfway between the parents in every respect.

Of course, such theories were ridiculous, and they did not explain why a family nose type would exist, why a child could so closely resemble one parent, or why some traits would skip a generation, showing up again in the grandchildren. After all, if each generation was a blend of the previous generation, it wouldn't be long before everyone would be more alike, and the species would eventually produce identical offspring.

At first, Mendel found that the result of pollinating a red-flowering four-o'clock (*Mirabilis jalapa*) with the pollen of a white-flowered plant would produce a first generation of pink flowers, results that supported the theory of blending. But in the second generation, things became more complicated. About one-quarter of the flowers produced from the crosses were white, one-quarter were red, and about 50 percent were pink.

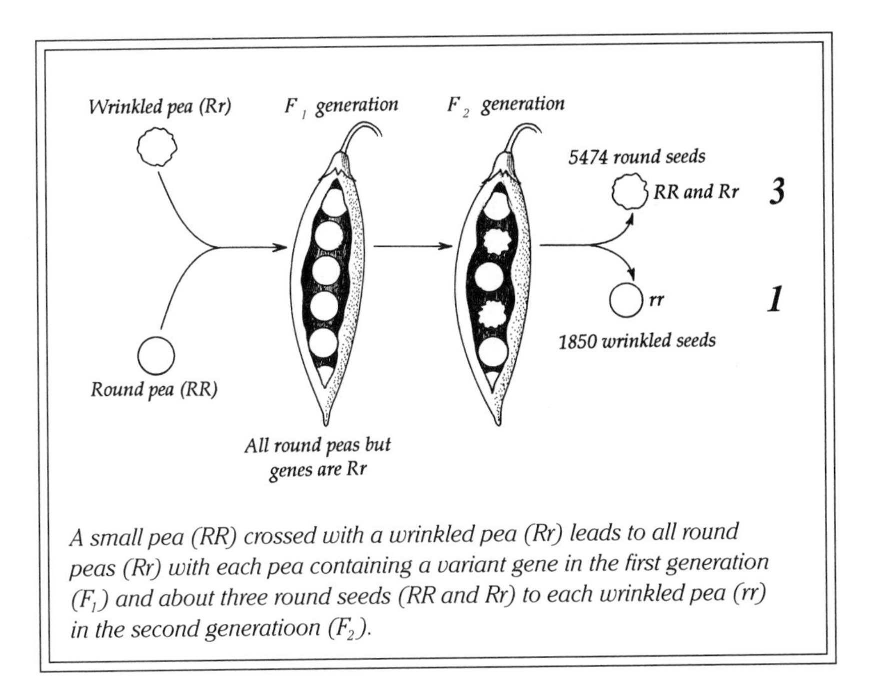

A small pea (RR) crossed with a wrinkled pea (Rr) leads to all round peas (Rr) with each pea containing a variant gene in the first generation (F_1) and about three round seeds (RR and Rr) to each wrinkled pea (rr) in the second generatioon (F_2).

After generations of experiments, Mendel found seven dominant traits: tall plants, yellow seeds, round seeds, purple flowers, flower position along the stem, pod color, and pod shape. The recessive traits were dwarf plants, green seeds, wrinkled seeds, white flowers, the flowers found at stem tips, yellow pods, and for shape, constricted pods.

But when he crossed peas from the first generation, the second generation produced a fixed ratio of color—one green pea to three yellow peas. One characteristic, here yellow, would dominate over the other color of green, but in succeeding generations, it was lost at times (or recessive) and reappeared at others. The drawing below shows a pea plant of the second generation. Its eleven pods have a total of forty-two yellow (here black) and fifteen green peas (here white), randomly distributed on the plant. By an exact count of the characteristics of the fruits found on a number of plants produced from crossbred seeds, Mendel always came back to a 3:1 ratio. The mix in individual pods may vary, but when all the peas of one plant were counted, they kept to the 3:1 ratio.

When Mendel started again, he chose peas with two characteristics, crossing a plant with round and yellow peas with another plant with wrinkled and green peas. This time the results worked out to be in a ratio of 9:3:3:1. He explained the results by pointing out that inheriting a seed color had nothing to do with the seed shape. With yellow or green peas that were either smooth and round or wrinkled, he found that the characteristics resulted in the same ratio. The

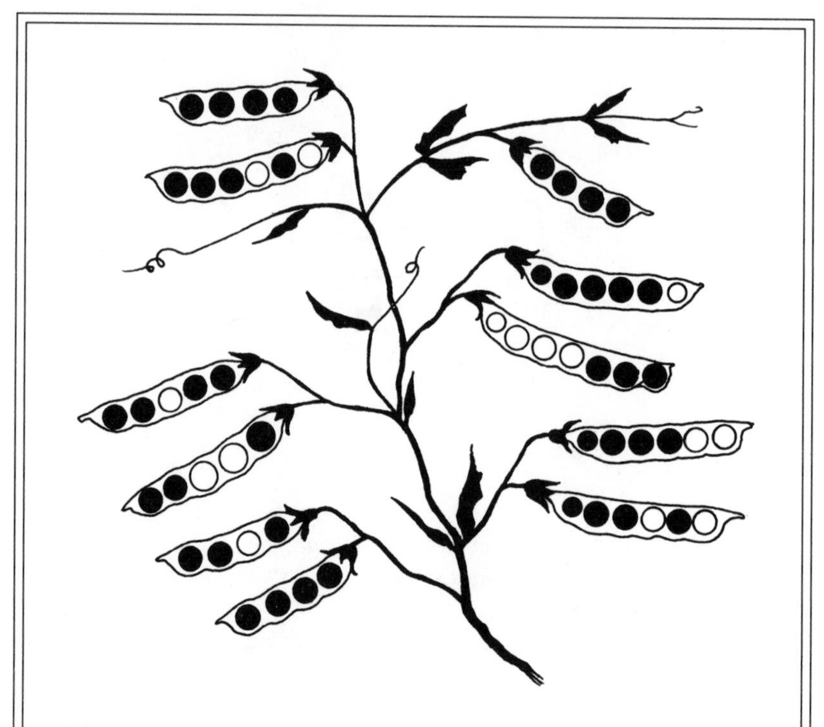

Mendel's tests with the garden pea. The sketch shows a pea plant of the second generation with eleven pods having a total of forty-two yellow peas (here black) and fifteen green peas (here white). The ratio is three to one.

probability of inheriting any combination of two traits can be calculated by multiplying the probability of inheriting each characteristic alone:

(round	**+**	***wrinkled)***	***x***	***(yellow***	**+**	***green)***
3	+	1		3	+	1
9	+	3		3	+	1
round		round		wrinkled		wrinkled
yellow		green		yellow		green

In reality, Mendel's work was pure genius. Before the discovery of chromosomes, genes, and meiosis, he discovered the basic laws of heredity.

In 1886, Hugo de Vries (1848–1935), a botanist from Haarlem in the Netherlands, found a species of evening primrose in an abandoned potato field. It was *Oenothera erythrosepala*, once called *O. lamarckiana*,

Or the way it's usually portrayed in what is called the Punnett square:

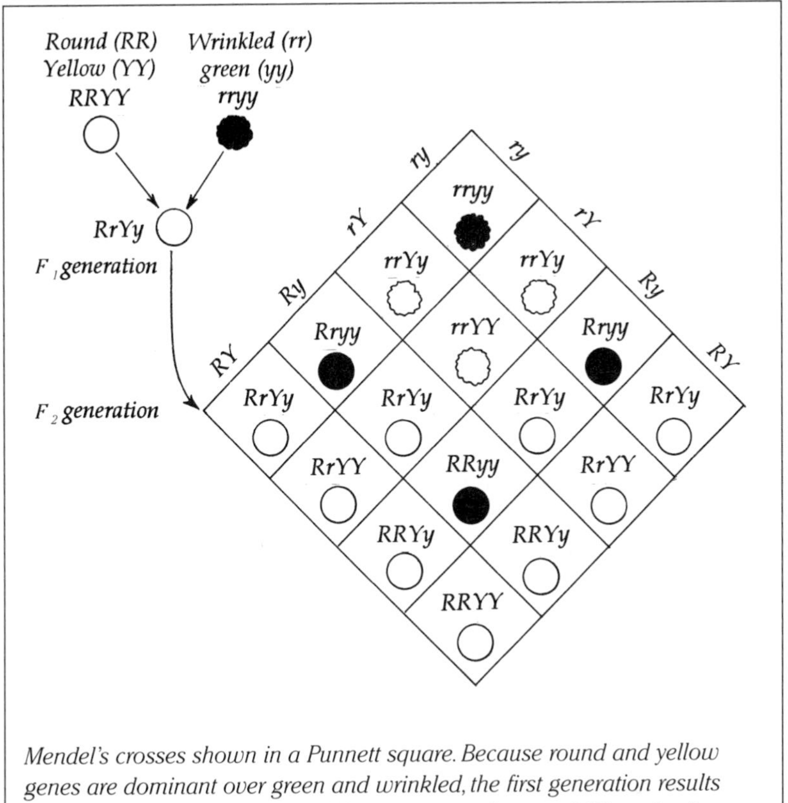

Mendel's crosses shown in a Punnett square. Because round and yellow genes are dominant over green and wrinkled, the first generation results in a round/yellow pea, but in addition to genes for round (R) and yellow (Y), the plant also carries genes for green (Y) and wrinkled (r). But in the second generation of peas, the results are nine round/yellow, three round/green, three wrinkled/yellow, and one wrinkled/green. All these ratios were predicted by Mendel.

a variety that arose in cultivation after the parent plant was introduced to Europe from somewhere in America. The original plant had produced several offspring in that field, and the progeny exhibited all sorts of flower and leaf forms, seed capsules, and other characteristics at odds with a so-called normal plant. Two of the plants were so different from the original stock that de Vries considered them to be a new species.

He took three plants back to the botanical gardens at Amsterdam. By carefully pollinating each plant with pollen from a plant of its own kind, he produced five thousand individuals over seven generations. More than eight hundred of them showed new characteristics or mutations. From the original *Oenothera erythrosepala* came a new dwarf form; a form that developed only sterile pollen, or none at all; a variety with long leaves; one with broad leaves and very large flowers; a form that he termed inconsistent; and a plant with many brittle stalks and broad red stripes on its calyx and its fruit. Nearly all these mutations from previous forms kept the new traits from generation to generation. And when de Vries crossbred the plants, the results behaved exactly like Mendelian laws would predict. Although he thought of them as being new species, it turned out they were all really stable hybrids. Without regard to his assumptions, de Vries had found a new answer to the question: How do new species originate?

F_1 AND F_2 HYBRIDS

Seed catalogs tend to make a big deal of F_1 and F_2 hybrids. And they often cost much more than seeds without that title. The easiest way to define an F_1 hybrid is to use the example of a plant breeder who spots a great color in one plant, but the plant has shoddy leaves. In a nearby plant of the same species, for example, he spots lackluster color combined with great-looking leaves. So the breeder takes the best plant of each type and self-pollinates them in isolation. The resulting seed is then resown until the breeder feels safe in saying he has a pure line, meaning every time seed is sown, identical plants appear.

Now, if the breeder takes one of each type from the pure line, and cross-pollinates the two by hand, the result is known as an F_1 hybrid. Plants grown from the resulting seed should exhibit good color and good leaf shape.

Complications often ensue, not the least that it can take up to eight years to produce a pure line. Sometimes, a pure line is produced after several previous crossings to ensure desirable features. Because so much extra work is involved, including a lot of hand labor, F_1 hybrids are expensive.

But remember Mendel's laws: Seed collected from an F_1 hybrid will not produce plants like the parents—except from a genetic point of view. Crossing F_1s to produce an F_2 generation will often produce a large percentage of acceptable plants, but some will revert to either showy colors with poor leaves or good leaves with poor color. F_2s will usually exhibit their parents' disease resistance and vigor. Also, they are cheaper seeds because hand-pollination is not needed.

Open-pollinated seed varieties are plants that are self-pollinating or cross-pollinated, lacking the uniformity of F_1 hybrids or most F_2s. If you find a good-quality snapdragon that is not billed as an F_1 or F_2, it's easy to save the seed, sow the seed next spring, and wind up with the same desirable plant.

BREEDING YOUR OWN FLOWERS

When breeding your own plants, bear a few things in mind: It's not an overnight process—it takes time. It requires close observation and a great degree of care.

Probably the best type of plants to begin with are annuals because they will bloom the first year, often within eight weeks of germination. For inspiration, think of the poppy or the sweet pea, and remember, the white marigold came from an amateur gardener.

If you want to ensure a particular characteristic is maintained, then self-pollination is the way. Germinate the resulting seed for a new round of plants, throwing out any misfits.

But if you wish to create new combinations of plant characteristics, you will want to cross-pollinate and must be careful to prevent self-fertilization. That means protecting the flowers from any chance pollen that bees or other insects might bring. It also means that any pollen-bearing parts of the flowers should be removed before unwanted pollen is shed.

Gardeners can make insect-proof cages for protecting plants in the open. I've used waxed-paper bags or plastic bags to prevent contamination, though I prefer paper to plastic.

Tools are simple. I use inexpensive camel hair watercolor brushes to transfer pollen from one flower to another, carefully cleaning the brush of pollen after use. A small pair of sharp tweezers is valuable for removing unwanted plant parts, and a good magnifying glass is necessary for close work. Finally, careful records are mandatory, and they must contain all the details of the crosses.

Once the seed starts to ripen, some fall directly to the ground; others fly away. Pods dry and burst. And some seeds are so small that if the container breaks, they are lost. So the plants must be watched every day, and the seed must be collected at the appropriate time.

When sowing your seeds, don't use up everything the first time. Things often go wrong, and it's best to have something up your sleeve for a second try. Then, grow as many plants as you have the room for—and the patience for. Remember, the characteristic you wish to produce might not show up in the next generation, so you will have to continue planting.

Some good annual flowers to try are alyssums, snapdragons, asters, annual morning glories, cosmos, lobelias, annual poppies, marigolds, and pansies.

Seeds of bottlebrush grass (Hystrix patula).

CHAPTER 4

Seed Chemistry

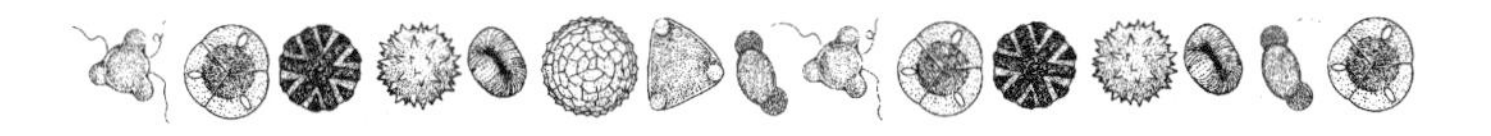

Majella Larochelle, a gentlemen of my acquaintance, has long been a seed collector. Everywhere he goes, the seeds are not far behind. A few years back, he gave me some paper packets containing seeds of a honeysuckle species (*Loniceria ledebourii*) from California, originally collected in England. One day in the greenhouse, working in haste as usual, I moved the packet and didn't notice that it accidentally fell behind a pile of catalogs on the work table. The catalogs and the seed packet sat there for two years. In that time the heat and light of the summer sun bleached the catalog colors from bright reds and yellows and greens to acid purples and off-blues, shriveling the paper in the process. The seed packet also had dried and shriveled; in fact, somewhere in recent history, it had been dotted with water. Majella's careful label was almost lost in grime.

This year I planted the seeds, and within a month, they germinated. Soon, I'll have a rarity—at least for this part of the world—honeysuckle in the garden.

Those honeysuckle seeds sat through all sorts of indignities, yet within each thin seed coat, living tissue managed to survive. The embryo, surrounded by the endosperm, put its life processes on hold. Within a state of suspended animation, the seed's contents still clung to life.

Seed storage proteins, deposited within special cellular organelles called protein bodies, stood side by side with glycoproteins, amino acids, neutral lipids, triacylglycerols (once called triglycerides), proteins, fats, carbohydrates, and other known (or perhaps to be discovered) substances—all in a little brown nuclei about an eighth of an inch wide. Under the direction of the nuclei of individual cells, simple chemicals were organized into complex chemical units, replete with energy, ready to be released when the proper time came.

THE SEED COAT

The seed coat, or testa, is the wall that stands between the seed embryo and the outside world. Considering how important the contents are, the package varies quite a bit. In fact, many seed coats are so individual in character, they have been used to differentiate between different genera and species.

Seed coats are often impregnated with fats or waxes that lend additional resilience and strength to their construction and that often lead to very effective waterproofing. In addition to the outer layer are one or more layers of thick-walled, protective cells.

The seed coats of water lilies, mallows, many legumes, and several morning glories have such tough seed coats that the seeds are virtually impermeable to water and must be nicked by the gardener or soaked in warm water for twenty-four hours before they germinate. If these jackets are not broken, scratched, or eroded, water never enters and germination never begins. Gardeners are a patient lot, but after a time, they'll attack such seeds with a file or soakings, where in nature, time and exposure to the elements does the trick. Hard seed coats are softened by alternately freezing and thawing or drying and wetting. Microorganisms in the soil can infect seed coats and will eventually rot them off.

The seeds of red and white clover (*Trifolium pratense* and *T. repens*) have a seed coat that is impermeable to water when the moisture content of the seed reaches a low level. There is actually a fissure along the groove of the hilum (the sear that marks the seed's attachment to the parent plant), which acts like a hygroscopic valve. When the seeds are surrounded by dry air, the fissure opens and allows water vapor to escape from the seed; the fissure, however, immediately closes in damp air, so the seeds can continue to dry without gaining any water.

Some seeds bear outgrowths of the hilum. One such structure is called the *strophiole,* and its function is to restrict water from entering or leaving a seed. Another type of structure is called an *aril.* Arils are appendages that vary in shape and, depending on the kind of seed, form ridges, bands, knobs, and even cup-shaped outgrowths of the testa. Some contain various chemicals; others may be brightly colored. Both attributes help to attract animals that aid in seed distribution. The aril of the nutmeg (*Myristica fragrans*) is the source of the spice called mace, whereas the seed coat itself is ground up to produce its namesake.

Seed coats often contain mucilaginous cells that break open upon contact with water and provide a barrier around the seed. The seed of the common plantain (*Plantago major*) becomes slimy the minute it hits water. And according to present medical authorities, such mucilage may lower cholesterol levels.

THE EMBRYO

The seed embryo of a dicotyledon is made up of the embryonic axis epicotyl (or the main body of the embryo) that bears two very large seed leaves or cotyledons. The axis itself is divided into the embryonic root (or radicle) at the bottom end, the central hypocotyl (or eventual stem), followed by the two cotyledons. Above the cotyledons is the hypocotyl, called the epicotyl, the part of the seedling that will eventually develop into the shoots. The very top of the epicotyl is known as the plumule and consists of several very tiny, immature leaves. It's all very straightforward, and all these parts are easily seen by examining a growing bean seedling.

But when it comes to monocotyledons, identification gets a bit more difficult. Unlike dicots, the monocot embryo stores its food in the surrounding endosperm, and the single cotyledon is modified and becomes the scutellum. A special structure called the coleoptile protects the young stems and leaves as they push up through the soil. The same job is done by the coleorhiza for the emerging root tip.

Just as there are twins in the animal world, a few species exhibit *polyembryony*, or more than one embryo in a seed. Twins occur for a number of complicated, developmental reasons, including a fertilized egg cell dividing into a number of parts. *Opuntia*, many citrus species, and the grass *Poa alpina* are examples of polyembryony.

THE NONEMBRYONIC STORAGE TISSUES

The endosperm is the food source for the embryo plant, and basically, it's also the food source for humanity. Remember, about 70 percent of all the food that people eat comes from seeds.

Seeds are divided into two categories: those with endosperms and those without. Some seed endosperms have a great deal of endosperm, like the cereal grains and the date palm, whereas others like lettuce have an endosperm that is only a few cell layers thick. Monocot seeds usually have a large endosperm compared with the embryo. The coconut has the most unusual endosperm, because although white and solid in a mature nut, it's in liquid form when immature.

If an endosperm fails to develop normally, the embryo will suffer. Such embryos can abort or permanently remain in an imperfect stage of development.

CARBOHYDRATES

Food reserves in a seed are mostly carbohydrates and starch. Other forms of carbohydrates occur as cellulose, pectins, and mucilages.

Starch is found in bodies called starch grains, and these grains have an individual aspect that is unique to a certain species. Such grains are elliptical in runner beans, angular in corn, and round in barley. In some seeds, the carbohydrates are stored as hemicellulose that appear as a very strong cell wall and make the endosperm extremely hard, like a coffee bean.

FATS AND OILS

Chemically, the fats and oils of seeds are known as triacylglycerols (once called triglycerides), usually in a liquid form above 68°C. In addition, some seeds will contain phopholipids, glycolipids, and sterols. The fatty acid found in corn and sunflower seeds is used to manufacture several cooking oils and margarine. The advantage, of course, to using these so-called vegetable oils is their higher unsaturated fatty-acid content.

PROTEINS

Among the many kinds of proteins worldwide, after cereal seeds, legumes are the most important source of proteins. Unfortunately, although cereal proteins are sufficient to provide adequate nutrition to human beings (if there are also enough calories), they are not enough for farm animals. And the amino acids found in both legume and cereal protein are not the best for either human or animal diets. But given their drawbacks, they are still amazing foods.

The different proteins have names like cysteine and methionine. There are holoproteins, acidic polypeptides, and a range of amino acids known as glycine, lysine, proline, serine, tyrosine, and valine, plus many, many more. Seed storage proteins are usually stored as special cellular organelles called protein bodies.

PHYTIN

Seed embryos need minerals too. Phytin is a mix of magnesium, potassium, and calcium. Some phytins also contain manganese, iron, copper, and rarely, sodium. Phytin is considered nutritionally undesirable because it can chemically bind essential dietary minerals like zinc, iron, and calcium, so they cannot be absorbed by human cells. How much of a problem this is in the Western world, where food processing removes most of the phytins, is not completely understood. In third world countries, where food is rarely refined, the presence of phytin may be a greater problem.

OTHER CONSTITUENTS

Seeds are made up of many chemicals. Alkaloids are organic, nitrogenous, basic substances derived from plants. They include morphine, codeine, strychnine, and quinine. There is theobromine in the cacao bean, caffeine in both coffee and cocoa. The soybean contains sitosterols and stimasterols, which are pharmaceutically important because they can be converted to the animal steroid progesterone. Presumably, nature prevents insects and animals from using seeds as food by including many of these substances.

Today, castor oil from castor beans (*Ricinus communus*) is rarely used for children's medicines. It is, however, a valuable ingredient in lubricants and waxes, not to mention a source of sebacic acid, used to make synthetic fibers and lubricants for jet engines.

Soybean oils are used in both plastics and resins. Copra comes from the oil-bearing flesh of the coconut. Other liquid fats from the oilseeds are used to make soap and glycerin for explosives. In tropical countries, solid fats are used to make candles. Several soft drinks contain extracts from kola nuts, the seed of the kola tree of the West Indies and South America. Zein, the alcohol-soluble protein from corn gluten, is used as an adhesive in printing inks. And for people who take a fiber supplement in their diets, a plant gum from psyllium seed (*Plantago psyllium*), once regarded as a nuisance plant, is used to promote regularity.

Germinating seedlings of an unknown lettuce.

CHAPTER 5

The Process of Germination

No matter where a dry seed sits quietly, it's in suspended animation, waiting for the one thing to force germination—water. But once water enters the seed, a transformation occurs: Metabolism begins, as well as respiration, enzyme activity, and RNA and protein synthesis.

Living things are able to duplicate themselves; inanimate objects cannot. To pass information from one organism to another, there must be a way to preserve information so that inherited characteristics can be passed on from the parent to the offspring. All these complicated instructions are encoded in the polymer deoxyribonucleic acid (DNA), the material used to make genes. In addition to DNA, there is ribonucleic acid (RNA), another organic chemical that acts like a template that can translate characteristics into a readable code in the cell. To call the process anything less than miraculous is to slight all the workings of nature.

Regardless of their age, many seeds will germinate the minute they absorb water. This reaction is especially true of crop seeds. Corn plants will soon appear after the seeds are planted in warm, moist soil. Other seeds, including many flowers and weeds, will not begin until special conditions are met, because some seeds have one or more blocks that delay germination (see page 47). If hellebore seeds germinated shortly after hitting the ground, the resulting seedlings would soon die in winter cold. Special blocking agents prevent the seeds from germinating until the next season.

But it's water that starts the process of germination, and it need not be a great deal of water. In most cases, it doesn't exceed two or three times the dry weight of the seed. Small seeds, seeds that have smooth coats, and seeds surrounded with mucilage are the best at absorbing water (that's why *Plantago psyllum* seeds, the main ingredient of all those fiber laxatives, thicken so fast when wet).

Water enters the seed, through either the seed coat or the tiny opening called the micropyle. Members of the papilionacea, including white sweet clover *(Melilotus alba)*, have a plug that must be removed before water enters the special opening called the strophiolar cleft.

The amount of time it takes for water to saturate the seed depends on the permeability of the seed coat, the seed's size, and the external environment, including temperature and humidity.

As the seeds soak up water, many chemicals are released, including sugars, organic acids, amino acids, and proteins. To prevent infections from bacteria or funguses in the soil, many seeds exude inhibitors as protection against invaders.

Experiments have shown that when pea embryos meet water, they leak potassium, electrolytes, sugar, and protein for about thirty minutes, but only from the outermost layers of the embryo.

In mature, dry seeds, the cell walls are shrunken and, when viewed under an electronic microscope, have a wavy or folded appearance like a serpentine wall. As soon as these walls are touched by water, they begin to expand and straighten out. After the internal process of germination is complete, the radicle, the part of the hypocotyl that eventually develops into the root system, pokes out of the seed and begins to force its way into the earth.

There is presently a debate in scientific circles about the radicle. To complete germination, the radicle must expand and push through the surrounding structures. Some botanists think the process happens because of cell expansion; others link it to cell division. Examinations of the broad bean (*Vicia faba*) show that radicle expansion occurs in two steps. During a first phase of forty-eight hours, about 30 percent of the seeds have radicles that have emerged from the seed coat. This phase is followed by a more rapid phase, accompanied by cell division in the radicle.

From experiments conducted on tomato seeds, a model has emerged to show the complexity of germination activities, including the roles played by the various hormones and water.

Mitochondria are the organelles that contain the biochemisty machinery responsible for metabolism. Experiments have shown that these structures begin to function during the first few hours that water enters a seed. Mitochondria in dry seeds lack definition and have no characteristic internal folds. But over a period of hours or days, they begin to function.

Shortly after water reaches the interior, food reserves are called up by the germinating seed. In seeds of foxglove (*Digitalis purpurea*), new

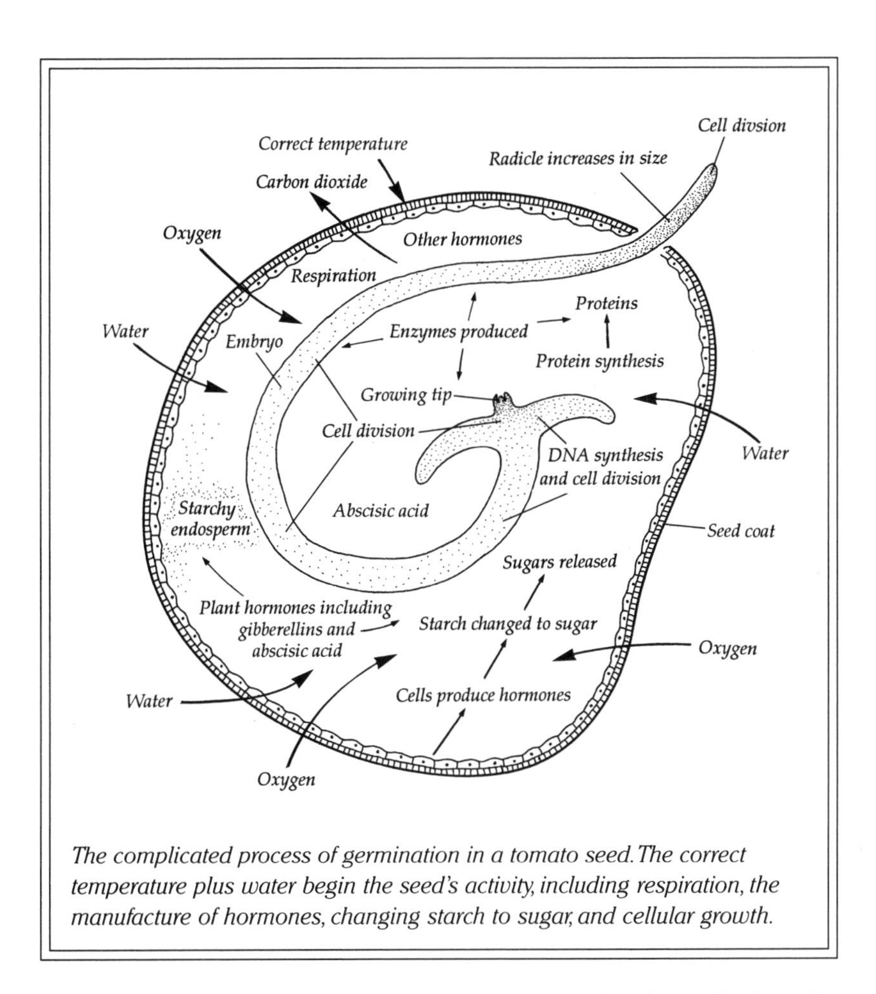

The complicated process of germination in a tomato seed. The correct temperature plus water begin the seed's activity, including respiration, the manufacture of hormones, changing starch to sugar, and cellular growth.

starch grains appear in the root cap at least twelve hours before the radicle begins to grow. Proteins and sugars are found both in the root tip and at the plumule tip.

The manufacture of protein begins within minutes of the water's reaching the seed's interior. Within a few hours, RNA and cell division begins, stored food reserves are released, and the radicle expands. To utilize these foods, enzymes must be produced for digestion and for future growth.

Dicot seedlings are divided into two types based on what happens to their cotyledons after germination. The broad bean is called *hypogeal* because the cotyledons remain beneath the soil while the root reaches up to the sky. A typical bean or squash seedling is called *epigeal* because here the cotyledons lift out of the soil, and the true leaves emerge from their protection.

Using a bean seedling as an example, let's follow a dry bean from its first contact with water to the point when the true leaves appear.

Germination in a Bean

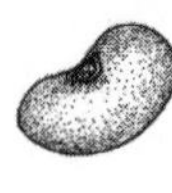

(A): A dry bean.
After the been absorbs water, its seed coat becomes wrinkled.

(B): The wrinkled seed.
The seed sliced open to show the embryo.

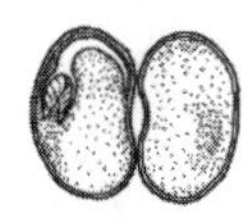

(C): The embryo.
The emergence of the radicle.

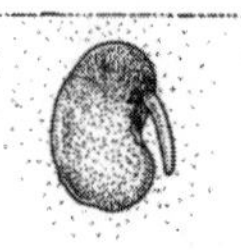

(D): The radicle.
The seedling pushes through the soil.

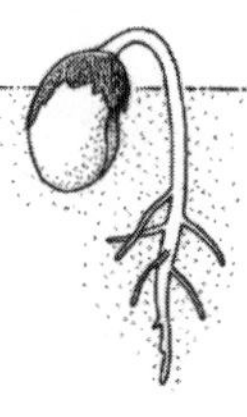

(E): The seedling pusing through the soil.
The emergence of the seedling with part of the seedcoat still clinging to the one on the right.

(F): The emergent seedling.
The seedlings begin to straighten up and the primary leaves begin unfolding. The plant on the right shows how the leaves fit together.

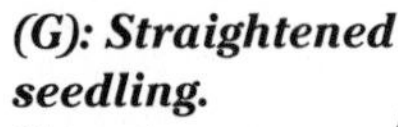

(G): Straightened seedling.
The primary leaves are open and the stem has elongated.

(H): Primary Leaves.
The true leaves have emerged.

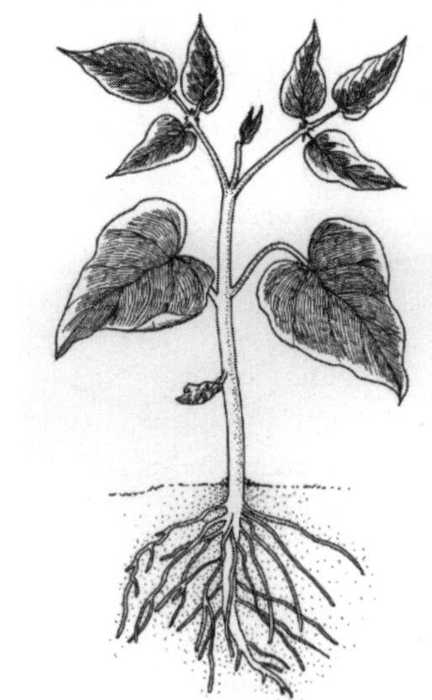

(I): True leaves.
Soon after the seedling is well above the earth, cell division begins in the plumule, the young growing tip of the stem.

MONOCOT SEEDLINGS

Corn, lilies, and grasses are typical monocotyledons and follow the same process of germination as dicots. However, because they lack cotyledons, monocots derive most of their initial nutrition from the surrounding endosperm in the seed. That's why, for example, the seeds of corn, wheat, rice, and barley are so nutritious. When a corn kernel germinates, the root tip breaks through a protective cover called the *coeleorhiza,* and the first leaf is protected on its journey through the soil by the *coleoptile.*

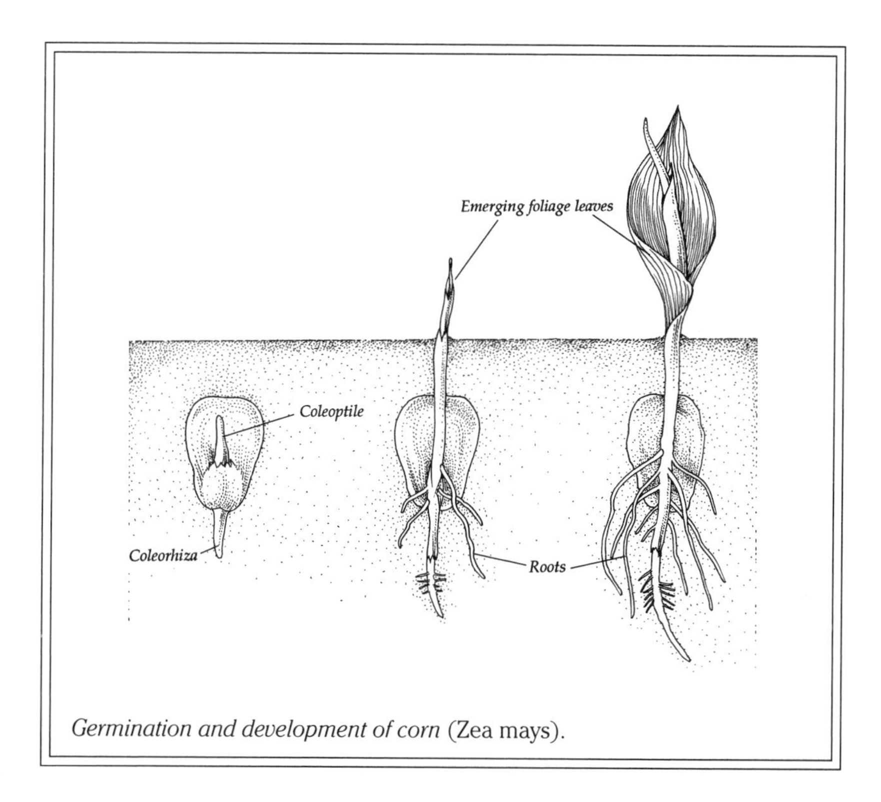

Germination and development of corn (Zea mays).

DAMAGE TO CHROMOSOMES AND DNA

Inside the seed coat, the embryo is a fragile bit of living tissue, in continual danger from any combination of time, moisture content, or temperature that proves to be more than its genetic material can withstand. Studies have shown that in several species, including the garden pea, breakages in the chromosomes appear in relationship to the length of storage and are evidenced by their appearance during the first mitotic divisions of the root tips. Insects, funguses, bacteria, chemicals, radiation, or light can diminish or destroy a seed's power to germinate.

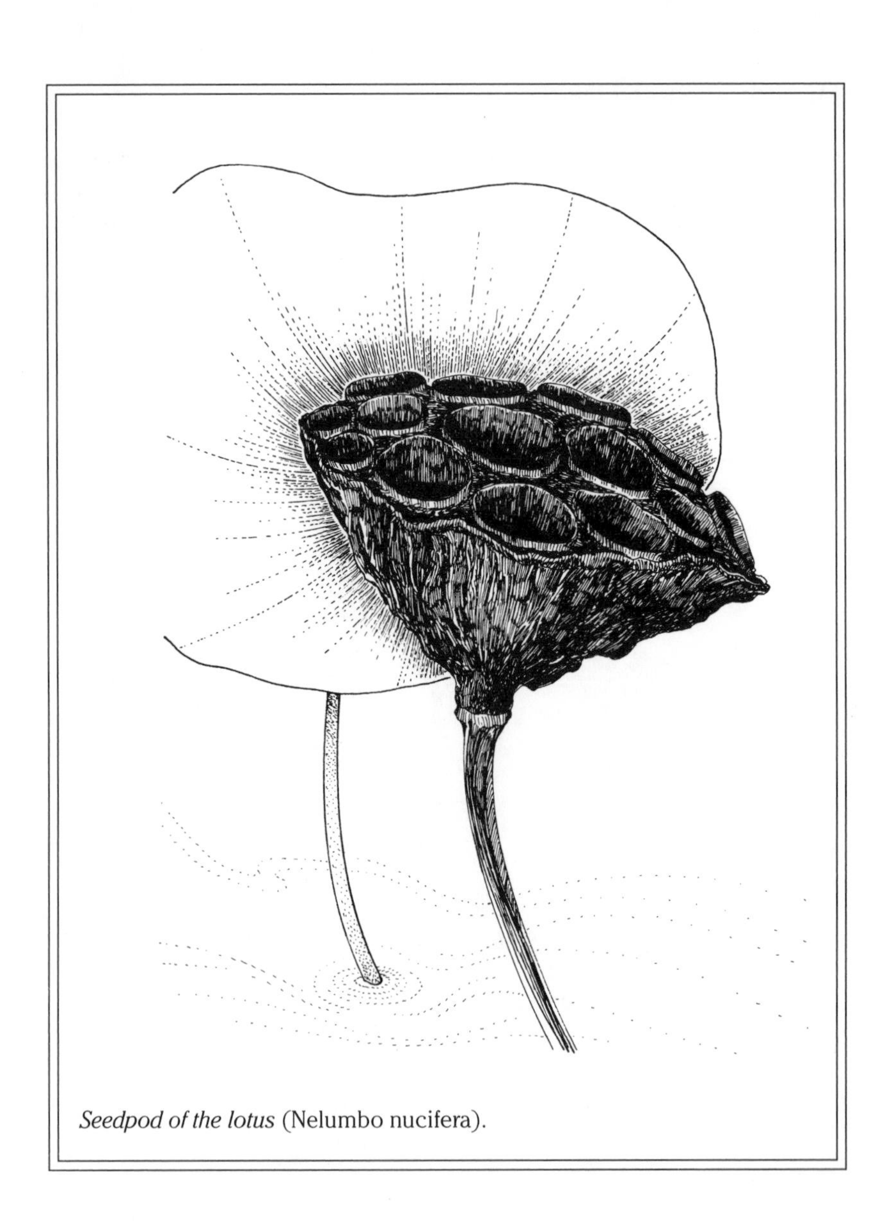

Seedpod of the lotus (Nelumbo nucifera).

CHAPTER 6

The Longevity of Seeds

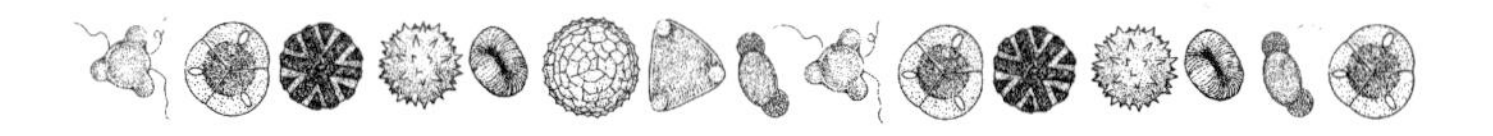

The following poem originally appeared in the December 1977 issue of the newsletter of the Henry Doubleday Research Association, an English organization devoted to the concept of organic gardening. Lawrence D. Hills, the author, wrote it in the style of Thomas Tusser (1524–80), an English writer known for his popular rhymed compendiums of advice for farmers.

> *You have in your drawer since Christmas Day,*
> *All the seed packets you daren't throw away.*
> *Seed Catalogues cometh as year it doth end.*
> *But look in ye drawer before money you spend,*
> *Throw out ye Parsnip, 'tis no good next year.*
> *And Scorzonera if there's any there,*
> *For these have a life that is gone with ye wynde.*
> *Unlike all seeds of ye cabbagy kinde,*
> *Broccoli, cauliflower, sprouts, cabbage and kale,*
> *Live long like a farmer who knoweth good ale.*
> *Three years for certain maybe five or four.*
> *To sow in their seasons they stay in ye drawer.*
> *Kohl-Rabi lasts with them and so does Pei-Tsai.*
> *The winter "cos-lettuce" to sow in July.*
> *But short is the life of ye Turnips and Swedes,*
> *Sow next year only, enough for your needs.*

Mustard and Cress for when salads come round.
Sows for three seasons so buy half a pound.
Radish lasts four years, both round ones and long.
Sow thinly and often, they're never too strong.
Last year's left lettuce sows three summers more.
And beetroot and spinach beet easily four.
But ordinary Spinach both prickly and round,
Hath one summer left before gaps in the ground.
Leeks sow three Aprils and one soon will pass.
And this is as long as a carrot will last.
Onion seed keeps till three years have flown by,
But sets are so easy and dodge onion-fly.
Store Marrows and Cucumbers, best when they're old.
Full seven summers' sowings a packet can hold.
Six hath ye Celery that needs a frost to taste.
So hath Celeriac before it goes to waste.
Broad Beans, French ones, Runners, sown in May,
Each hath a sowing left before you throw away.
And store Peas, tall Peas, fast ones and slow,
Parsley and Salsify have one more spring to sow.
Then fillen ye form that your seedsmen doth send.
For novelties plentie, there's money to spend.
Good seed and good horses are worth the expense,
So pay them your dollars as I paid my cents.

Obviously, when it comes to vegetables, there has probably been a tremendous interest in the longevity of seeds. And if you are worried only about keeping popular vegetable seeds, following the instructions cited in the poem will work admirably well.

Even centuries ago, records point to an interest in just how long seeds will live. A review of the available literature points out that the Greek philosopher Theophrastus (c. 372–c. 287 B.C.) mentions that "of seeds some have more vitality than others as to keeping," and "in Cappadocia [an ancient region of Asia Minor] at a place called Petra, they say that seed remains even for forty years fertile and fit for sowing." Unfortunately, until recent years, most of the research conducted on seed longevity dealt with those seeds found in botanical gardens, usually stuck in dusty bins and cupboards, crammed in crumbling brown paper envelopes, or those falling from dried seed pods on mounted herbarium specimens.

In 1908, in his treatise *On the Longevity of Seeds*, A. J. Ewart noted that presumably no area of human knowledge documented more misleading, incorrect, or contradictory statements than the literature dealing with seed longevity (obviously, if he had been writing today, he would have traded seeds for politics). In an attempt at establishing order, Mr. Ewart divided seeds into three biological classes according to their life under favorable conditions. Microbiotic seeds had a life span not exceeding three years, mesobiotic seeds could live from three to fifteen years, and macrobiotic seeds retained viability from fifteen to one hundred years. Under the macrobiotic classification, he listed 137 species in 47 genera of the family Leguminosae (peas); 15 species in 8 genera of Malvaceae (mallows); 14 species in 4 genera of Myrtaceae (myrtles); 3 species in 2 genera of Irideae (iris); and 1 species in each of the families Euphorbiaceae (spurges), Geraniaceae (geraniums), Goodeniaceae (goodenias), Polygonaceae (buckwheat), Sterculiaceae (sterculias), and Tiliaceae (linden or basswoods).

In her monograph on *Seed Preservation and Longevity* (Leonard Hill [Books] Ltd., London, 1961), Lela V. Barton of the Boyce Thompson Institute for Plant Research reported the following:

> *One of the earliest records of germination of old seeds is that of A. de Candolle (1846). Seeds of 368 species representing various families were collected, principally from the botanical garden of Florence in 1831, and were tested in 1846 when nearly fifteen years old. They had been kept in a cabinet protected from high humidity and extremes of temperature. Only seventeen of these species germinated when the seeds were planted in soil, and only one of these,* Dolichos unguiculatus, *showed more than 50 percent germination. Leguminosae (5 out of 10 species tried) and Malvaceae (9 out of 45 species tried) accounted for 14 of the 17 viable lots of seeds. The seeds tested represented 180 annuals, 28 biennials, 105 perennials, and 44 woody plants, together with 11 unidentified seed lots. The author concluded that the figures obtained seemed to prove that woody species preserve their viability longer than others, and the biennials, none of which were found to be viable, deteriorated most rapidly.*

In 1907, Paul Becquerel, a French botanist, tested five hundred seeds found in the seed collection of the Museum of Natural History in Paris. Records showed the age of these seeds to be between 25 and 136 years old. Four families, Leguminosae, Nymphaeaceae (water lilies), Malvaceae, and Labiatae (mints) produced germinating seeds, with twenty of the seeds ranging from 28 to 87 years old. The seeds of *Cassia multijuga* (a Brazilian tree) germinated after 158 years of storage, an apparent record at the time. The following table shows the results of his tests.

Macrobiotic spp.	**Date 1906; 1934**	**% Germinated:**	**Years of longevity: real—probable**
Mimosa glomerata	1853	50—50	81—221
Melilotus lutea	1851	30—0	55
Astragalus massiliensis	1848	0—10	86—100
Cytisus austriacus	1843	10—0	63
Lavatera pseudo-olbia	1842	20—0	64
Dioclea pauciflora	1841	10—0	93—121
Ervum lens	1841	10—0	65
Trifolium arvense	1838	20—0	68
Leucaena leucocephala	1835	20—30	99—155
Stachys nepetifolia	1829	10—0	77
Cassia bicapsularis	1819	30—40	115—199
Cassia multijuga	1776	100	158

Mr. N. G. Moe, the head gardener of the Botanical Museum of the University of Oslo, collected seeds from 1857 to 1892 and stored the seeds in strong paper bags. He wrote the name of the species and the date on the outside. Another collection at that garden was started by A. G. Blytt, who gathered seeds during the first years of his professorship, 1880 to 1898, storing them in corked bottles. Germination tests were conducted during the years 1932–33, so the ages of the seeds ranged from approximately 34 to 112 years. In all, 1,254 different groups of seeds were tested, and out of these, only 53 species showed any germination. The oldest living seeds were those of a species of milk vetch (*Astragalus utriger*), at 82 years, with 6 percent germination.

Seed from the silk tree or mimosa tree (*Albizzia julibrissin*), now naturalized in the southeastern United States, was collected in China in 1793. Deposited in the British Museum, London, the seed germinated after attempts to put out a fire started by an incendiary bomb that hit the museum in 1940.

Others who have experimented with seed longevity include Mr. J. H. Turner of Kew Gardens, who in 1933, reported viable seeds in samples of seven species of legumes that were eighty or more years of age. The genera were *Anthyllis, Cytissus, Lotus, Medicago, Melilotus,* and *Trifolium.*

Finally, in 1948, Fritz W. Went of the California Institute of Technology and Philip A. Munz of Rancho Santa Ana Botanic Garden in California started an elaborate longevity test on seeds of more than one hundred native California plants. Samples were dried in vacuum desiccators, then packed in small glass tubes to 0.1 millimeter of mercury or less, sealed, and placed in an insulated but unrefrigerated storage room. Supposedly, the last set of these samples will be tested for germination in 2307 A.D.

SEEDS BURIED IN THE EARTH

Seeds that are stored inside dry cabinets or sealed in unopened tombs of arid Egypt represent one problem, but seeds that have been stored underground present an entirely different enigma. Henry David Thoreau worked on the manuscript for *The Dispersion of Seeds* from 1860 to his death in 1862. The finished work finally appeared in the book *Faith in a Seed* (Island Press, Washington, D.C., 1933). He wrote the following about the vitality of seeds:

> *I am prepared to believe that some seeds, especially small ones, may retain their vitality for centuries under favorable circumstances. In the spring of 1859 the old Hunt House, so called, in this town, whose chimney bore the date 1703, was taken down. This stood on land which belonged to John Winthrop, the first governor of Massachusetts, and a part of the house was evidently much older than the above date and belonged to the Winthrop family. For many years I have ransacked this neighborhood for plants, and I consider myself familiar with its productions. Thinking of the seeds which are said to be sometimes dug up at an unusual depth in the earth, and thus to reproduce long-extinct plants, it occured to me last fall that some new or rare plants might have sprung up in the cellar of this house, which had been covered from the light so long. Searching there on the 22d of September, I found, among other rank weeds, a species of nettle (*Urtica urens*) that I had not found before; dill, which I had not seen growing spontaneously; the Jerusalem oak (*Chenopodium botrys*), which I had seen wild in but one place; black night shade (*Solanum nigrum*), which is quite rare hereabouts; and common tobacco, which, though it was often cultivated here in the last century, has for fifty years been an unknown plant in this town and a few months before this, not even I had heard that one man, in the north part of town, was cuiltivating a few plants for his own use. I have no doubt that some or all of these plants sprang from seeds which had long been buried under or about that house, and that tobacco is an additional evidence that the plant was formerly cultivated here. The cellar has been filled up this year, and four of those plants, including the tobacco, are now again extinct in that locality.*

Seventeen years later, the longest experiment to determine the life of seeds buried in soil was conducted by W. J. Beal. Mr. Beal welcomed in the year 1879 by mixing seeds of twenty different wild species with sand and burying them approximately eighteen inches below the surface. Mr. Beal used uncorked pint bottles, but their mouths were turned down to prevent their filling with water. Testing was accomplished by transferring the contents of the storage bottle to a seed flat filled with sterilized soil, then placing it in a greenhouse.

After forty years of burial, seeds of eight species were still alive. These survivors were pigweed or redroot amaranthus (*Amaranthus retroflexus*), ragweed (*Ambrosia elatior*), black mustard (*Brassica nigra*), peppergrass (*Lepidium virginicum*), the biennial evening primrose (*Oenothera biennis*), common plantain (*Plantago major*), purslane (*Portulaca oleracea*), and sour-dock (*Rumex crispus*). All produced seedlings, but after seventy years, only *Oenothera biennis* and *Rumex crispus* would germinate. Because I find the biennial evening primrose worthy of inclusion in any garden of interest, I was delighted with its pluck. But sadly, sour-dock is well named, and unless it was one of ten plants left in an otherwise barren world, it could never pass for anything but a pernicious weed.

The Seed Testing Laboratory of the United States Department of Agrilculture started such testing in 1902, terminating the experiment in 1946. Seeds of 107 species, representing both cultivated and wild plants, were buried in thirty-two sets in flower pots filled with sterile soil and capped with porous clay covers. They were buried at depths of eight, twenty-two, and forty-two inches. Germination tests were reported after burial for 1, 3, 6, 10, 16, 20, and 39 years. Seventy-one species germinated after one year, 61 after three years, 68 after six years, 68 after ten years, 51 after sixteen years, 51 after twenty years, 44 after thirty years, and 36 after thirty-nine years. The highest germination rate came from the seeds buried at the forty-two-inch depth, and the lowest from those seeds at the eight-inch depth.

After those thirty-nine years, the sixteen species with the highest germination rates were as follows:

velvet leaf	*(Abutilon theophrasti)*
ragweed	*(Ambrosia artemisiifolia)*
bindweed	*(Calystegia sepium [Convolvulus sepium])*
jimsonweed	*(Datura stramonium)*
morning glory	*(Ipomoea lacunosa)*
bush clover	*(Lespedeza intermedia)*
tobacco	*(Nicotiana tabacum)*
evening primrose	*(Oenothera biennis)*
Scotch thistle	*(Onopordum acanthium)*
pokeweed	*(Phytolacca americana)*
annual cinquefoil	*(Potentilla norvegica)*

black locust	*(Robinia pseudoacacia)*
black-eyed Susan	*(Rudbeckia hirta)*
common nightshade	*(Solanum nigrum)*
red clover	*(Trifolium pratense)*
common mullein	*(Verbascum thapsus)*

Again, the results of this experiment will prove of no major news value to dedicated gardeners who know about the perseverance of these plants. Because of a personal liking for black-eyed Susans, evening primroses, and pokeweeds, I would not call them weeds. But for centuries these weeds have also been known as "the seeds of a troubled earth," usually appearing whenever dirt is disturbed after long years of lying fallow.

According to Dr. Barton, a most remarkable characteristic of the longevity of weed seeds, especially when stored in the soil, is that most of them lack impermeable seed coats. Therefore, these seeds immediately absorb water when they are exposed to moist soil. The fact that the seeds, after soaking up their full measure of water, would remain viable over years is quite remarkable. Apparently, these seeds do not develop a "deep" dormancy, because when they are disturbed in a garden plot, they immediately begin to germinate—even if care has been taken to remove all such plants for several preceding years. Exposure to light, alternating temperatures, mechanical disturbance, or some other unknown reason just might be the stimulus that shocks these seeds to life.

SEEDS FROM EGYPTIAN TOMBS AND OTHER ANCIENTS

Right up there with the mummy's curse that supposedly led to the death of Lord Carnarvon, the archaeologist who uncovered King Tutankhamen's tomb (more popularily known as King Tut), are stories of wheat seeds taken from mummy bindings that germinate after thousands of years. (It was really a tomb fungus that did the explorer in; that's why today's archaeologists wear filter masks when entering ancient sealed tombs.) In the 1860s, Thoreau mentioned the stories of wheat raised from seed that was buried with an ancient Egyptian tomb and of raspberries raised from seeds found in the stomach of an Englishman who was supposed to have died sixteen or seventeen hundred years before Thoreau's time, though he generally discredited these stories because of the lack of evidence.

Recent scholarship has proven that most of these seeds have undergone severe morphological and physiological degradation along with total loss of viability. Although many of these seeds evidence a cell structure, when water is added, a great deal of disintegration occurs. Barley seeds at an age of about 3,350 years that were collected from Tut's tomb were extensively carbonized and completely nonviable.

To date, the oldest claim for longevity is for the arctic lupin (*Lupinus arcticus*), after seeds were found frozen and buried in the Canadian Yukon. The seeds were taken from an ancient rodent burrow that contained the skull of a lemming. Carbon dating of the nests and remains of Arctic ground squirrels found buried under similar conditions in central Alaska registered an age in excess of ten thousand years. The researchers concluded that the seeds in the Yukon were also of this age. Even a gardener would consider such reasoning wildly optimistic. Knowing the perfidies of moles and that they are carnivores, I wouldn't be surprised if an enterprising Canadian member of the genus stopped to rest in this ancient tunnel and brushed off some seeds that were stuck to its fur.

As to other ancient seeds, between 1843 and 1855, Robert Brown found 150-year-old viable seeds of the sacred Indian lotus (*Nelumbo nucifera*). In 1925, further testing revealed that they had lost the power to germinate over the intervening seventy years. But during the 1920s, seeds of that species of lotus, which had been collected in a layer of peat from a naturally drained lake bed in southern Manchuria, germinated. Radiocarbon tests indicated their age at 1,040 plus 210 years old, making them some of the oldest viable seeds ever found. Notes taken at the time show that these seeds would not germinate without treating the hard seed coat to allow water permeation. This was achieved by either dipping the seed in concentrated sulphuric acid or scratching it with a file.

Another case of the Indian lotus concerns seeds found on a three-thousand-year-old submerged boat at Kemigawa, near Tokyo. The seeds proved to be viable, but no direct measurements were made of the seeds' age, and it's possible they could have settled in the mud after falling from modern plants.

Seeds of lamb's-quarters or pigweed (*Chenopodium album*) and seeds of toadflax or corn spurry (*Spergula arvenis*) were found at an archaeological dig in Denmark and were thought to be about seventeen hundred years old; but no direct dating was used, and they could have been modern weed seeds that had infiltrated the dig. Most gardeners would likely believe they were weeds, because weed seeds seem to live forever.

Other seeds collected from ancient times include the species *Canna compacta*, found in an Argentinian tomb and encased in the nutshell of a species of walnut, *Jugland australis*. The walnut was part of a rattle necklace, proved by carbon dating to be about six hundred years old. After germination there was not enough of the seed for analysis, but the only way it could have gotten inside the walnut was to have been inserted while the developing nutshell was still soft.

THE IMPORTANCE OF THE SEED COAT

The seed coat, or testa, is extremely important for seed longevity. After all, the seed coat is all that stands between the embryo and the outside world, although in some cases, the fruit coat or even the endosperm can be a substitute for this structure. Usually, there is an outer and inner cuticle, often impregnated with waxes or fats, followed by one or more layers of thick-walled, protective cells, a palisade, or Malpighian layer made of heavy-walled, tightly packed, radially placed, columnar cells that are not only hard, but also horny. There are no intercellular spaces allowing water to accidentally enter the interior.

Many seeds are protected by layers of cells that contain calcium oxalate or calcium carbonate crystals, which protect seed embryos not only from the weather, but also from insects. Some seed coats contain mucilaginous cells that burst when touched with water but provide a water-retaining barrier of protection.

When the seed detaches from the parent plant, it bears a scar called the hilum (see page 17). At one end of the hilum, many species of seeds bear a small hole called the *micropyleremains.* The micro-pyleremains will stay plugged until germination occurs.

PLANTS FROM OLD SEEDS

Years ago, there were occasional reports about seeds that were better after a certain amount of time had passed. An Irish *Farmer's Gazette* from the mid-1800s stated "the gardener knows that melon and cucumber seeds, if used of the last year's saving, produce plants too vigorous to produce much good fruit; whereas, those kept over for several years produce less rambling, but very fruitful plants." An anonymous gardener in 1932 talked about the relative values of pumpkin seeds, claiming that in one trial, three-year-old seeds gave the best results, whereas in another trial, four-year-olds yielded less quantity but a far higher quality fruit.

But most of the experimental evidence shows that older seed, unless properly stored, produces inferior plants.

STORING SEEDS

When it comes to storing seeds, outside of using futuristic chromium tubes, surrounded by helium under pressure, or any other supermodern methods of keeping, what should the gardener do?

First, except for a very small class of seeds known as recalcitrant seeds, which quickly deteriorate when they lose water, most seeds are at their best when stored in a dry state. The most common recalcitrant seeds are the willows (*Salix* spp.), hazel (*Corylus avellana*), black walnut (*Juglans nigra*), coconut (*Cocos nucifera*), coffee (*Coffea arabica*), and wild rice (*Zizania aquatica*). In general, there are few guidelines for recalcitrant seeds and much research waiting to be conducted. For example, wild rice is recalcitrant only when dried at temperatures below 25°F and then rehydrated, or provided with water. But seeds will survive if dried at temperatures above 25°F and watered slowly. If you have any of these seeds, my advice is to plant them immediately, and do not confine them to storage.

Naturally, the seeds that prefer dry conditions are known as orthodox seeds.

For orthodox seeds, remember to follow these general rules:

1. For each 1 percent decrease in seed moisture, the storage life of the seed is doubled.

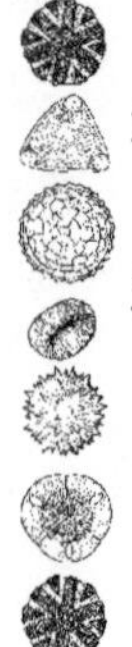

2. For each 10°F (5.6°C) decrease in seed storage temperature, the storage life of the seed is doubled.
3. The sum of the storage temperature in degrees Fahrenheit and the relative humidity should not exceed 100 percent, with no more than half the sum representing the temperature. So if the storage temperature is 40°F, then the relative humidity should not be more than 60 percent; if the temperature is 70°F, the relative humidity should be below 30 percent.

OTHER FACTORS IN STORING SEEDS

Other than temperature, moisture content, and the passage of time, some other problems can surface when storing seeds.

The first problem begins in the field where seeds are gathered. Differences can be found according to the cultivar and the harvests. There might be a defect in one plant that is not noticeable in the garden but will show up later in the seed. Also, the ripening process could be interrupted during harvest, although many seeds will continue to mature even when in storage. The differences are usually minor, and they usually do not show up in seeds that have been properly stored. But when storage considerations are not the best, that's the time for weaknesses to show. In other words, some seeds have better potential than others.

Although bacterial damage is not likely when seeds are properly stored, occasionally, insects get into a seed collection and begin to chow down. This problem usually occurs only in countries with high temperatures and high humidity and improper storage.

So keep your seeds in a clean, dry place with low temperatures and low relative humidity. Unless I have been sidetracked in the potting shed or greenhouse, I keep seeds in their original packages or if self-collected, store them in glassine stamp envelopes. Then I put them in a sealed baggie and keep them near the floor of the potting shed or the greenhouse, where even in the summertime, the temperatures are cool and the air is dry.

If I must keep seed that is valuable to me, I will place a small packet of silica gel, the kind that comes packed with cameras, in the envelope to keep away excess moisture. Doc and Katy Abraham, who have been growing and saving seeds for decades, suggest that seed savers wrap some dry milk in a bit of paper tissue or toilet paper, secure the packet with a rubber band, and place it in the plastic bag along with the seeds.

Seedpod of a flowering-maple (Abutilon *spp.*).

CHAPTER 7

Seed Dispersal

Anyone who has ever had to brush a longhaired dog after its late autumn tramp through woods and fields knows how some seeds travel. Seeds become affixed to pants, skirts, and jackets, using hooks, barbs, gluelike material, and even natural systems of velcro to guarantee their dispersal around the country if not the world. Seeds travel in many ways in addition to hitching a ride on or in an animal.

DISPERSAL BY WIND

Small seeds often use the wind to blow them from place to place. Some are like dust, flying through the air until eventually falling to earth. One of the smallest seeds is a pernicious and parasitic weed called the witchweed (*Striga asiatica*). The tiny seeds are 0.0078 of an inch long, and each plant produces hundreds of thousands of seeds that, because of their size, can literally blow around for miles.

The seeds of most orchids are as fine as dust. Because it is so difficult for an orchid seed to find the right place to germinate and grow to maturity, each flower must produce a prodigious amount of seeds. A scientific count of a capsule of a *Cycnoches chlorochilon* reached the fantastic total of 3,770,000.

"The seed of cattleyas is as fine as powder," wrote Rebecca Northern in *Home Orchid Growing* (Prentice Hall, New York, 1990), "and close to a million are formed in one capsule. Its small size allows it to be carried by the gentle air currents that are not strong enough to lift heavy particles to great heights in the jungle. As the seed capsule ripens and splits open, the drift of pale yellow powder is picked up by moving air and dusted from branch to branch."

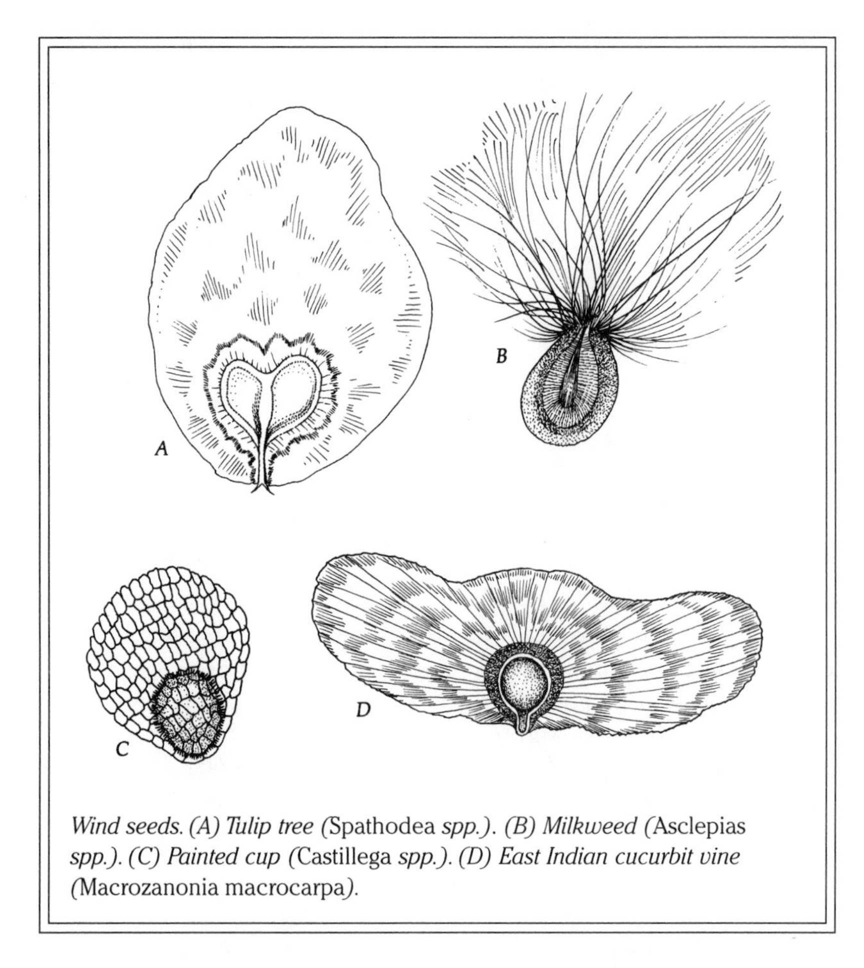

*Wind seeds. (A) Tulip tree (*Spathodea *spp.). (B) Milkweed (*Asclepias *spp.). (C) Painted cup (*Castillega *spp.). (D) East Indian cucurbit vine (*Macrozanonia macrocarpa*).*

When Hurricane Hugo flew up along the southern coast of the United States, it brought along a large number of various seeds of tropical plants on the trip. Sara Peacock, a neighbor of mine whose mother lives in Charleston, South Carolina, brought me seeds from three unknown plants that sprang up in her mother's backyard shortly after the storm.

"Drifting along with the tumbling, tumbleweed," goes the song that tells the story of this unusual seed spreader. When ripe and dry, tumbleweed (*Amaranthus albus*) plants are torn loose from the soil, and as they blow along, they drop seeds as they pass.

Plumed seeds like fireweed (*Epilobium*), epiphitic bromeliads like the tree-dwelling Tillisandias, dogbane (*Apocynum*), thistle seeds, and dandelions (*Taraxacum officinale*) use their silken plumes to fly through the air like a badminton bird.

In every garden I've ever had, there has been either the giant field milkweed (*Asclepias syriaca*) or the smaller, but still attractive butterflyweed (*A. tuberosa*). If you have room for such a wild plant, the first milkweed is worth growing because of its attraction to the monarch butterfly, while the second attracts butterflies of all descriptions. But being milkweeds, they both produce large pods that are stuffed with hundreds of seeds, each seed attached to a plume of white silk. As the pod dries, it splits, and the plumes emerge seed first, each waving about in the summer air. With a final tug at the pod, the plumes fly off like the characters in the waltz of the flowers from the movie *Fantasia*. The plumes of the common milkweed were also gathered by children during the Second World War for use as stuffing in what were cheerfully called Mae West life preservers for use by the armed forces.

Some fruits that contain seeds actually have wings that enable them to glide through the air. Maples (*Acer*), ash (*Fraxinus*), elms (*Ulmus*), and dock weeds (*Rumex*) all have seed coats that develop into thin wings. The airplane seed produced by an East Indian cucurbit vine (*Macrozanonia macrocarpa*) has a three-foot wing on either side of the seed and turns in a spiral about twenty feet wide as it falls to the ground.

Many plants bear seeds that are not winged but are so flat and membranous that they are easily carried aloft by the wind. Tulip seeds (*Tulipa*), for example, have narrow marginal wings, and I've watched a gentle breeze hustle them along the flat of the garden. Both lily and yucca (*Yucca*) seeds are flat and packed in a seed capsule like a deck of cards. As the capsule splits, the seeds emerge a few at a time and are blown long distances before they fall to the ground.

Most of the rockcresses (*Arabis*) have long, narrow pods that release elliptical seeds, each bearing a marginal wing for easy flight in a breeze. The horseradish-tree (*Moringa pterygosperma)* is grown throughout India and bears curious three-angled, winged seeds that are discharged from a large ribbed pod about fifteen inches long, ready to float on the winds. The seeds are collected for a valuable oil.

Gardeners who have watched the spread of the beautiful flowering tree called the paulownia, or empress tree (*Paulownia tomentosa*), know its prolific nature. Its leathery, ovoid capsules contain hundreds of small, delicate seeds, each of which has two or three wings with the capability to cover miles in search of a place to land and germinate.

Various seeds have a seed coat that is covered with woolly hairs, for example, willows and poplar (*Salicaceae*), cotton (*Gossypium*), kapok (*Bombax*), and anemones (*Anemone*).

The wind also disperses the seeds of many of the grasses, not because of any special adaptation, but because the seeds are light and aerodynamically adapted to flying in the wind.

Long before I read Thoureau's journals, I knew about white pines and white pine seeds. The autumn winds would toss the pine branches about with such ferocity that the cones would often fly many feet from the tree that grew them. Invariably, the cones would be empty. The oval seeds bear one long wing, and when the lid of each seed chamber lifts, the seed quickly blows away in a stiff wind or gently twirls to the ground if becalmed.

All winter long, some of the cones remain high on the branches, but if you wait to collect seed from a fallen cone, you will have no luck. Sometimes you will find the remains of a white pine cone looking like a six- to eight-inch corn cob that is covered with brownish-red bristles. This is all that's left after the squirrels have completed their work. And often in late summer and early autumn, you will find immature cones, still green and oozing pitch, but still devoid of seed because all the small birds and animals have done their job of eating. Yet there are still plenty of white pines, because nature usually, but not always, makes sure there are more than enough seeds to guarantee the survival of each species.

DISPERSAL BY WATER

Whether by rain, streams and creeks, sudds, or ocean currents, many seeds are distributed by water. Very small seeds, especially if they are light in proportion to their size, will easily float. Corky seeds, like those of the carrot family, will stay afloat for weeks. The thick-walled fruits of the silverweed (*Potentilla anserina*) have been recorded as floating in streams for fifteen months. Thus the weed is distributed along riverbanks and swampy meadows throughout the northern hemisphere.

The beautiful-flowered but deadly jimsonweed (*Datura stramonium*) has flat, corky seeds that easily float. Throughout their range, the seeds are carried downstream by floodwaters and deposited on riverbanks or along the edges of roads, where they germinate. Many sedges (*Carex*) have seed pods that contain air pockets, so they easily stay on top of the water, sometimes for months.

Marsh marigolds (*Caltha palustris*) and several aquatic plants or those adapted to wet places have corky seeds. The marigolds never seem to make pests of themselves, but the loosestrifes (*Lythrum salicaria*) are particularly difficult.

In the tropics, especially in areas of heavy rain, the periodic rush of water from the mountains is awesome. This occurrence transports millions of seeds from the heights to germinate in the plains below. Most of the time, these seed migrations are governed entirely by chance, but a few plants have adapted to these conditions, and their seeds use water for a definite purpose. The pearlwort (*Sagina*) and the mitrewort (*Mitella*) have open cuplike capsules that contain seeds. Splashing raindrops dislodge the seeds, and they wash out to the earth below.

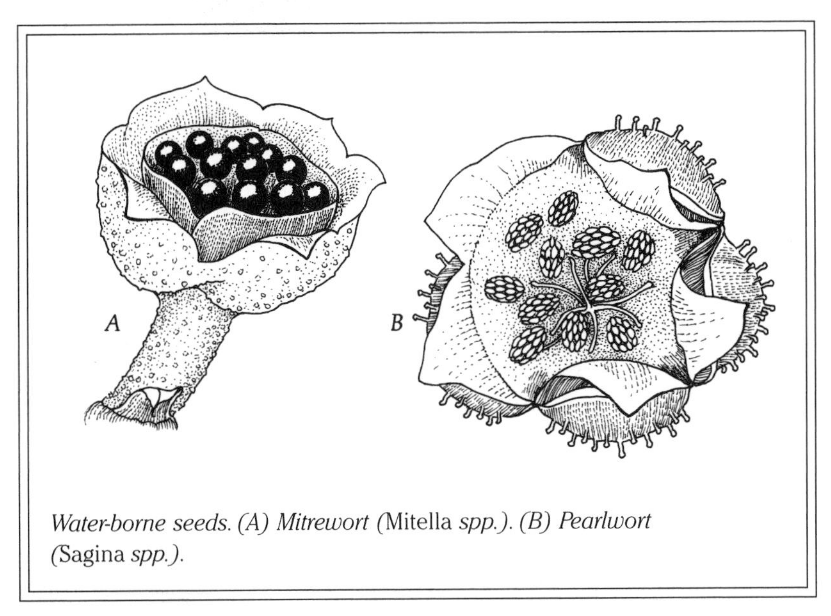

*Water-borne seeds. (A) Mitrewort (*Mitella *spp.). (B) Pearlwort (*Sagina *spp.).*

One of the Chinese water chestnuts (*Trapa bicornis*) has blackish fruits about three inches across that look exactly like a bull's head with two stout, curved sharp-pointed horns. These fruits can float great distances before sinking to the bottom of a shallow stream, where they germinate.

SUDD

Sudd comes from the Arabic word, *sadd*, for barrier, and it originally stood for the vegetation that made the White Nile impossible to navigate. Today, it refers to masses of dense vegetation that block water channels. A whole mass of plants, along with the seeds, can be torn away by a flood and carried downstream, usually ending in adjacent lakes or pools. Examples of sudd plants are papyrus (*Cyperus papyrus*), water chestnut (*Trapa*), water lettuce (*Pistia*), and the infamous water hyacinth (*Eichornia*) that has plugged so many rivers in the Deep South.

DISPERSAL BY THE OCEAN

For beachcombers of all ages, there is a fascinating book called *World Guide to Tropical Drift Seeds and Fruits* by Charles R. Gunn and John V. Dennis (Quadrangle, 1976). Within its pages, hundreds of plant species that use the ocean currents to distribute their seeds, often for thousands of miles, are described in detail.

Known as disseminules or sea-beans, tropical drift seeds and fruits all have the capacity to drift for at least one month in seawater. Viability is never a concern for the collector because some floating seeds were never alive, some lost their vitality while drifting, and others are stranded in good condition, and immediately sprout.

Seeds that float are divided into five groups:

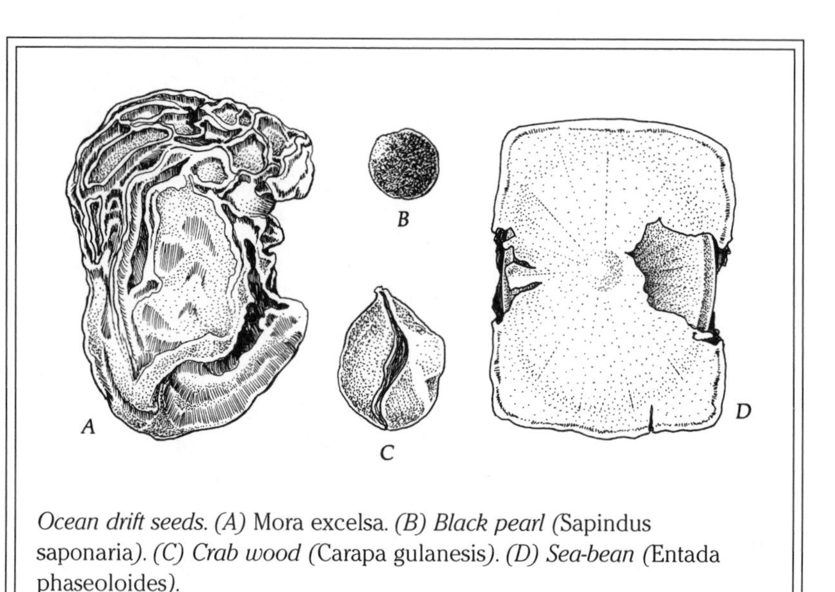

Ocean drift seeds. (A) Mora excelsa. *(B) Black pearl (*Sapindus saponaria*). (C) Crab wood (*Carapa gulanesis*). (D) Sea-bean (*Entada phaseoloides*).*

Group 1. These seeds have buoyancy because of a hollow cavity and usually because the endosperm does not completely fill the seed. A prime example of this group is the sea heart (*Entada gigas*), a pantropic, high-climbing woody vine. Its seeds have reached Norway, have buoyancy for at least two years, and are usually viable.

Group 2. These seeds float because of a lightweight tissue. The bay-bean (*Canavalia rosea*) has been found in most tropical currents and remains impermeable to water for at least eighteen months. Found on tropic beaches, the seed germinates into a ubiquitous strand vine that forms large tangles.

Group 3. This group is buoyant because of a fibrous or corky coat or a combination of both. An example is the country-almond or Indian-almond (*Terminalia catappa*). This native of tropical Asia has been spread by man, by bats, and by ocean currents. The seeds float because of a very soft and

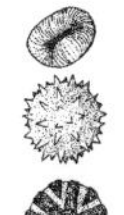

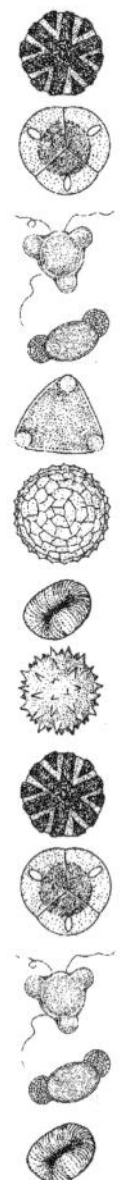

buoyant seed coat, retaining buoyancy for at least two years, with more than half the collected seeds being viable. The seeds produce excellent shade trees and have the taste of almonds.

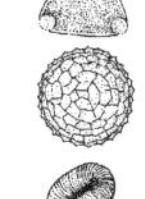

Group 4. This group is buoyant because of the thinness of the seed. Here, the best example is the yellow flamboyant tree (*Peltophorum inerme*), a native of the East Indies and widely planted throughout the tropics for its colorful flowers and usefulness as a fine shade tree. The seeds will float for at least nine months, and most seeds found are viable.

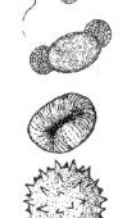

Group 5. Buoyancy in this group is due to a combination of all the above factors. Coconuts, the fruit of the coconut palm (*Cocos nucifera*), are the second largest of all known seeds and a perfect example of this group. Although it's true that the winner for size is the so-called double coconut (*Lodoicea maldivica*), no stranded seed has ever been found to be viable. The true coconut is ranked as one of the ten most important tree species, and although the authors of the cited book have never tested the coconut for viability, germinating stranded and drifting coconuts have been observed many times.

In the annals of drifting seeds, I cannot overlook the black mangrove (*Avicennia germinans*). The seeds belong to group 4, but unlike most disseminules, the black mangrove usually drifts as a seedling, not as a seed. The embryo germinates while the fruit is still attached to the parent tree. When the seedling drops, it either roots in the mud below the parent tree or is carried out to sea on the tide. There is a wonderful story that the unfolded cotyledons serve as miniature boats, but floating is correctly ascribed to the buoyancy of the seedlings themselves.

DISPERSAL BY ANIMALS

Earlier, I mentioned the seeds that are spread by animals, including people who go for long walks through the woods and fields. These transportations may be either external or internal, depending on how often the seeds are used for food.

DISPERSAL THROUGH INGESTION

Everyone who gardens in areas where poison ivy is found continually wonders why, year after year, the ivy comes back, when every last vestige of this pest was removed the year before. The answer is birds eat poison ivy seeds, then expel them as they fly around the yard. There's a crevice in the stone wall next to my front door where a chickadee inserts one sunflower seed every year as a hedge against tomorrow—and I'm sure others do the same around the yard.

Birds and mammals, even turtles, snakes, and tortoises, eat a number of fleshy fruits, then pass the seeds through their alimentary tracts unharmed. And in many cases, the germination rate speeds up after the journey. Raspberries (*Rubus*), Oregon grapes (*Mahonia*), and flowering dogwoods (*Cornus florida*) are all spread after being eaten by chipmunks, mice, birds, squirrels, and all the other inhabitants of a suburban or country backyard. Large seeds or small seeds like grasses can survive the digestive trip.

Many freshwater fish feed on vegetation, including the seeds of water plants, carrying them over long distances before they are expelled. But the most curious manifestation of this method was mentioned by Darwin in *Origin of Species.* Herons and other birds, he wrote, have eaten fish in whose stomachs were viable seeds of the yellow water lily. In this manner, the herons often carried the seeds many miles from their source.

The large garden snails of California and England have been known to move strawberry seeds throughout their range. And in southern climates, land crabs will eat fallen fruits and are known to spread the seed of the Malayan Otaheite chestnut (*Inocarpus edulis*).

The missel thrush has long been associated with the distribution of the European mistletoe (*Viscum album*) because the birds eat the berries. The berry's flesh contains a number of glutinous materials that are only partially digested by the bird's alimentary tract. When passed from the bird, the seeds and skins of the fruit are stuck together, and the mass frequently is fastened securely to the branches of trees. Other observers have seen mistletoe seeds stuck to the beaks of birds, who rub against bark to remove them, thus putting them in position for germination.

Even the lowly earthworm has been found, upon dissection, to have a gut containing a wide variety of seeds.

PURPOSELY STICKING TO FUR AND FEATHER

Spring in my garden is not only evidenced by daffodils and crocuses, but also by sprouting oak trees. While foraging for acorns and other nuts, the squirrels moved their food from one end of their territory to another, then forgot where they left them. The ground in and around our bird feeders is always sprouting sunflowers, millet, and various grasses resulting from the seeds thrown about by the birds.

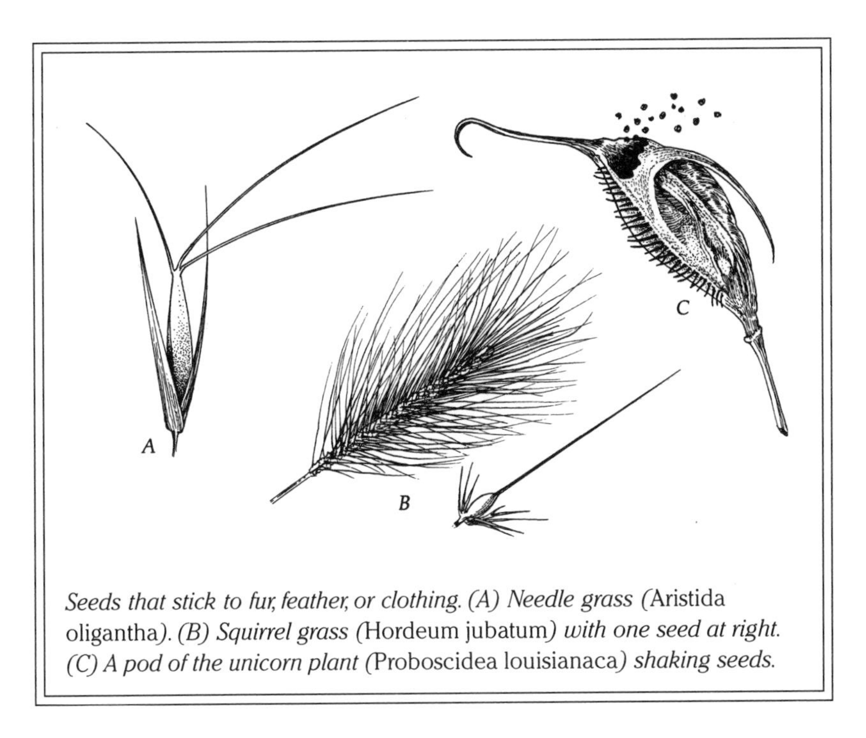

*Seeds that stick to fur, feather, or clothing. (A) Needle grass (*Aristida oligantha*). (B) Squirrel grass (*Hordeum jubatum*) with one seed at right. (C) A pod of the unicorn plant (*Proboscidea louisianaca*) shaking seeds.*

One spring, after gathering some seeds of hound's tongue or beggar's-lice (*Cynoglossum* spp.), Thoreau spent a long time removing them from his pocket. The following August, he gave the seeds to a young lady who cultivated a flower garden. He wrote the following about the results.

> *[The lady's] expectations were excited and kept on the qui vive for a long time, for it does not blossom till the second year. The flower and peculiar odor were sufficiently admired in due time; but suddenly a great hue and cry reached my ears, on account of its seeds adhering to the clothes of those who frequented these gardens. I learned that that young lady's mother, who one day took a turn in the garden in order to pluck a nosegay, just before setting out on a journey, found that she had carried a surprising quantity of this seed to Boston on her dress, without knowing it for the flowers that invite you to look at and pluck them have designs on you and the railroad company charged nothing for freight. So this plant is in a fair way to be dispersed, and my purpose is accomplished. I shall not need to trouble myself further about it.*

The numbers of adaptations that allow seeds to stick to animals are legion. Sometimes the entire seed capsule is involved as in the burdock (*Arctium*), where the various projections are hooked at the tip. Many members of the Compositaes, or daisy clan, have developed barbed awns that easily stick to fur or feather. The barbed bristles of the beggarticks (*Bidens*) are perfectly adapted to stick to fur or most other materials. The small flat pots of Tick-trefoils (*Desmodium*) are covered with hooked hairs. Both plants get their common names from sticking like ticks, and I'm sure were the inspiration for velcro.

As beautiful as it is, did you ever wonder how Queen Anne's lace (*Daucus carota* var. *carota*) travels so well around the garden and fields? The seeds bear hooked bristles and easily attach themselves to any rough surface.

Many seed coats have adapted to travel by having a gluey surface that will stick to almost anything. Seeds of the butterfly-pea (*Clitoria mariana*) are so viscid they will adhere to any passing animal.

This group of plants includes the plantains (*Plantago*), as well as a group of weedy plants hated by people who love a perfect lawn because they seem to infect any green sward of grass. Like many of the rushes (*Juncus*), some members of the mustard family, including the common garden cress (*Lepidium sativum*) and the shepherd's-purse (*Capsella bursa-pastoris*)—both imported from Europe—become viscid when wet. Some of the flaxes (*Linum*), and even several genera of the phlox family, have seeds that, when wetted, emit mucilate in the form of fine threads. The process has two separate advantages: Either the seeds blow about until they reach a damp spot and then germinate or they begin their travels in damp weather and still alight when it's best for germination to begin. Often, these sticky seeds affix themselves to dried leaves and are blown about until such time they find a place to grow.

Many grass seeds have awns or stiff hairs that actually twist and turn during periods of high humidity. The twisted portion of the awn of Porcupine grass (*Stipa spartea*) coils and uncoils as the moisture content of the air changes, causing the bent arm of the awn to revolve slowly until it touches grass stems or other objects. Then, the whole seed is literally screwed down into the earth. Unfortunately, the same process occurs if the florets lodge in the wool or hair of animals, often causing serious puncture wounds to grazing animals, usually around the eyes, nose, and mouth.

But when it comes to grabbing hold, nothing beats the fruits of the unicorn flower or ram's horn (*Proboscidea louisianica*), an American annual from the South. The fruit lies upon the ground with its six-inch claws sticking up into the air. When an animal steps on it, the fruit tips up and the claws clasp the fetlock. Then seeds are shed as the animal moves about.

From South Africa comes the grapple plant (*Harpagophytum*), this time with woody fruits that have four wings, each cut into a number of very stout linear arms with strong hooks on the tips. Again, the hooks become attached to unfortunate animals, and as they struggle for release, the seeds fly about with ease.

INSECT CARRIERS

I've watched a few beetles and a great many ants walking through the garden carrying seeds. One researcher has reported that fruits and seeds of more than sixty genera are carried about locally by ants, sometimes at a distance of more than one hundred feet. The seeds of hellebores (*Helleborus*) are attractive to ants, who unintentionally sow them, protecting them from the rigors of winter until the following spring, when they germinate.

The harvesting ants (*Messor*) of the Mediterranean region and down to the Sahara build subterranean nests that go down ten feet or more into the ground and are topped with a dome about two feet high. The ants follow well-worn trails and go out foraging for seeds that are used to make a paste that is fed to larvae. The seeds are stored in underground granaries, but if it rains and the seeds get wet, the ants take them out to dry in the sun, where some germinate.

MECHANICAL DISPERSAL

Just like the wheat that is shot from guns in order to puff up and float in milk, there are several seeds that depend on fruit explosions to disperse their seeds.

The plant that usually comes to mind when dealing with shooting seed is the squirting cucumber (*Ecballium elaterium*), an annual vine with fruit that is really an oblong berry about two inches long. It is native to the Mediterranean region, and Linneaus coined the genus *Ecballium*, which in Greek means "to throw," "eject," or "cast out."

The plant has a loose trailing or prostrate habit. When the fruit is ripe, it is a bluish green, about three inches long, and hangs by a stem (or peduncle) from the vine. Suddenly, the fruit will separate from the peduncle while the fluid tension within the fruit forces the outer layers to distend under pressure. From the scar, the fluid contents—basically a semiliquid mucilage—push out through the opening, carrying the seeds with them. This ejaculation often leads to the discomfort of the collector. While not entirely attractive, the squirting cucumber certainly fills the bill as a horticultural oddity.

Another plant from the gourd garden that explodes when ripe is *Cyclanthera brachystachya* (at one time bearing the species name of *Explodens*). The spiny, soft fruit with an up-curved end literally breaks apart at maturity, shooting out dark brown or blackish seeds about three-eighths of an inch across.

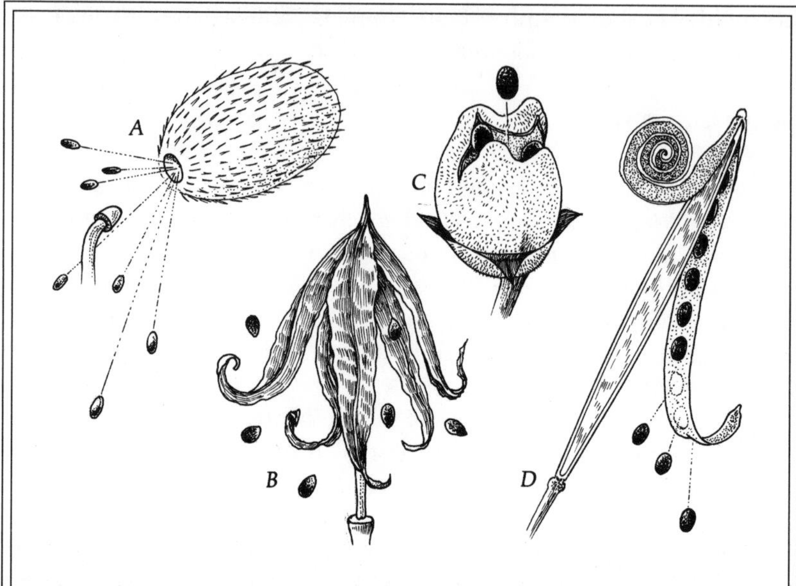

*Seeds that use mechanical aids. (A) Squirting cucumber (*Ecballium elaterium*). (B) Impatiens (*Impatiens aurella*). (C) Witch hazel (*Hamamelis virginiana*). (D)* Cardamine hirsuta.

Some popular garden plants also use force to spread their seeds. One is the popular touch-me-not or snapweed, belonging to the genus *Impatiens.* The fleshy seed capsules are often dilated at the upper end where the seeds are borne. The walls of the capsule are elastic, and as the seeds mature, pressure builds within. Soon, the five individual sections (or valves) are held together with only the slightest bit of tissue. Eventually, they either part on their own or react to the touch of another plant, animal, bird, or human. Either way, in the blink of an eye, each of the five sections rolls back like a party favor, throwing out the seeds to the open air.

Gardeners with greenhouses are usually familiar with the yellow-flowered creeping oxalis (*Oxalis corniculata*), a weedy inhabitant of damp floors. Using the same principle as the impatiens, the seeds are shot to the open air and quickly germinate in any favorable spot.

Many species of the geranium (*Geranium*) have developed a seed distribution system based on releasing tension. Each fruit has five carpels containing one seed each. The carpels are all part of a long thin column fused together at the top. As the seeds ripen, the column begins to separate into five fibers. As tension builds, the columns split apart and wind up to the top like five springs. The seeds are thrown as far as ten feet.

Several members of the legume family distribute their seeds when the fruits dry and split apart into separated spirals, sending the seeds more than eight feet. The violet-flowered Chinese wisteria (*Wisteria chinensis*) has hard, rounded seeds within a stout pod. When the pods open, seeds can fly ten feet or more.

The record for legume-flying seeds is probably held by the West Indian swordbean (*Canavalia gladiata*). When the large, thick pods snap open with a crack, the seeds are often thrown a distance of twenty feet.

Unfortunately, for many gardeners, one of the weeds that came along in a boat ballast is the spring vetch or tare (*Vicia sativa*), another member of the legumes that employ exploding seed pods to send seeds traveling up to ten feet from the parent plant.

In witch hazels (*Hamamelis*), the fruit is a woody capsule with two sections or valves. When the fruit dries, it splits apart, revealing two shiny black seeds within each valve. Then as the fruit continues to dry, it finally snaps apart and can discharge the seeds as far as forty-five feet, a necessary distance to fall outside the perimeters of the parent tree.

The small mistletoe (*Loranthaceae pusillum*) is a parasite of the black spruce and lives in the woods of the Northeast. When the berries ripen, in early fall, the seeds are violently expelled. The seeds have a mucilaginous surface, allowing them to stick to other parts of the bark on either the host or a nearby tree.

The sandbox tree (*Hura crepitans*), or the monkey-pistol, is a tropical tree found in the West Indies, Costa Rica, and Central America. The seeds are contained in round, three-inch capsules that upon ripening, will break apart into many sections and emit a sound that is heard throughout the jungle.

DISPERSAL BY HUMANS

The final way that seeds travel around the world is by the fine hand of man. For millennia, people have traveled and carried their foodstuffs with them—including seeds of all kinds. These seeds have clung to clothing, hidden in pants cuffs, or wound up in mud stuck to shoes.

Again in *Origin of Species*, Darwin reported an examination and study he made of about half a pound of dried mud. He wet it and kept it covered under conditions best for germination. Within six months, no less than 537 seedlings of various species appeared. In another case he took a ball of mud from the leg of a partridge and from this raised eighty-four plants of three species.

But when it comes to America, one of the easiest ways that immigrant seeds came to our shores, especially unwanted weed species, is in the ballast of sailing ships. Those round-bottomed boats that sailed from Europe and England left with empty holds. After all, few people in the colonies had the money to buy much of anything. Unfortunately, when empty, the ships would easily capsize, especially in storms. So for the voyage across the sea, shipping companies filled the hulls with dirt, then hired men to dig it out after landing. Guess where the dirt went? All along the eastern shores from Boston to Charleston. And all that dirt was full of seeds, hence, the alien weeds that now trouble our ecology.

CHAPTER 8

New Thoughts on Seed Germination

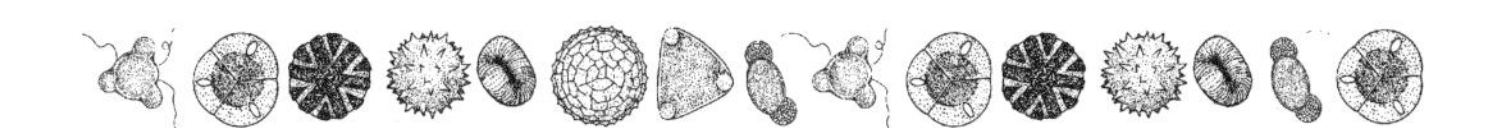

"Every species in the plant kingdom must have a mechanism for preventing seed germination before the seed is scattered." The statement is direct, the kind of assertion that could have been made in the late eighteenth century by a follower of Gregor Mendel. It came from Norman C. Deno when I met him at a wildflower conference in Cullowee, North Carolina, in July 1994, and he had full reason for saying it.

Norman C. Deno, a large man with short-cropped gray hair, began his career as a chemist. Although he did not become deeply involved with the problems of seed germination until after his retirement in 1980, he had been growing plants from seed since he was ten years old. At fifteen, he gave a talk before a garden club in Wilmette, Illinois, on growing lilies from seed.

Mr. Deno worked for twelve years to write (and self-publish in August 1991) *Seed Germination Theory and Practice*:

> *The first purpose of my book was to identify and to analyze these mechanisms of delay using my studies on the germination of 2,500 species of temperate zone plants that represent 123 families and 605 genera.*
>
> *The second purpose was to present an extensive set of specific directions for the optimum germination of the 4,000 species that had been studied to date. Miraculously, specific treatments for effecting good germination were found for nearly evey one of the 4,000 species studied, although the treatments are often complex. Three failures were blue cohosh* (Caulophyllum thalictroides), *bloodroot* (Sanguinaria

> canadensis*), and American holly (*Ilex opaca*). Curiously, all three of these plants self-sow in my gardens. For the first two it's suspected that they require a gibberellin for germination, other than the gibberellic acid-3 used in the experiments.* Rhinanthus minor *from Iceland has also persistently failed to germinate. This is a parasitic plant, and it may require a specific chemical from the host plant for germination, as has been shown for Striga.*

When I asked him what was his inspiration for the book, without a pause he mentioned the *Seedlist Handbook* by Bernard E. Harkness, first published in 1974. It was the result of Harkness's interest in alpine gardening and the thousands of phone calls he received during his twelve-year stint as seed director for the American Rock Garden Society.

STRATIFICATION

Most of the books written about starting plants from seed have a section called stratification, the process of exposing seeds to varying temperatures to break their dormancy and snap them to germinating attention. The word *stratify* refers to placing something in layers. The old horticultural definition in *Webster's New International Dictionary* (1925) reads: "The placing of seeds between layers of sand or sawdust, common in the case of seeds requiring to be kept moist or frozen to preserve their vitality and to facilitate germination, as those of many trees."

"Not so," said Mr. Deno.

> *The term breaking dormancy sounds like the seeds have been resting or were asleep. My interpretation is that these cold periods were actually periods of maximum metabolic activity for the destruction of chemical systems that were blocking germination. This proposition is based on a first principle of physical organic chemistry, namely that if a prior treatment is necessary for the chemical process called germination to occur, some kind of chemical reaction has to be taking place prior to the event. There are, of course, truly dormant conditions, but they are unnecessary for later germination.*

Mr. Deno describes the necessary treatments that occur before germination begins as chemical processes that destroy germination inhibitors, all part of a necessary chemical process. And one of the most remarkable things he discovered was that inhibitor destructions took place at temperatures of either 40°F or 70°F. With only one or two exceptions, no species were found where that activity took place at both those temperatures.

"It's like a rubber band," he said.

> *A rubber band can sit in a drawer for a year or more without any apparent change in its outward appearance or any change in its elasticity. Then one day a crack appears in the rubber, the band ruptures, and ultimately crumbles. The explanation is that raw rubber is rapidly oxidized by the oxygen in the air and a complex chain reaction results in the cleavage of the polymeric chains and the rubber crumbles to powder. When the rubber band is manufactured, chemical inhibitors are added to block the chain reaction and preserve the rubber. Ultimately these inhibitors are used up allowing the oxidative degradation of the rubber to begin.*

He goes on to discuss the various physical mechanisms that inhibit germination. For example, many legumes use thick and hardened seed coats to prevent germination before the seed is scattered about. Mr. Deno is not sure whether it is a lack of oxygen or water or both that holds up germination. But it is clear that grinding a hole in the seed coat will speed germination greatly. Instead of taking from one to six months, or even years, with low germination rates, the treated seeds show 100 percent germination, usually in fewer than ten days, and often in two to three. Some palms and cycads have a semipervious seed coat that delays germination. Making holes in the seed coat speeds germination within these groups.

GERMINATION

Germination, is when the seed begins to develop. Plant physiologists usually begin recording at the point when respiration or chemical activity first begins, while a gardener's record usually starts when either the seed leaves, cotyledons or the first true leaves appear above the ground. In my studies, the process was clocked from the moment the radicle (or embryo root) or rarely the epicotyl (the upper regions of a plant including the growing tip [see page 16]), could be seen breaking through the seed coat. Because the radicle elongates quickly—either by cell elongation or cell division—there is really no problem in deciding when a seed had germinated.

But interesting things do happen. Some species of Corydalis, Daphne, and Euonymus split the seed coat, expanding about 50 percent, and then stop. The development of the radicle takes place much later. Many Cornus and Prunus species expand enough in a cold cycle to cause their hard seed shells to split and even fall off leaving the radicle to develop later when temperatures are warm.

In a number of species, there is a lengthy time gap between the development of the radicle and the time when the true leaves emerge. Sometimes a period of two to three months is needed, in which the temperatures are kept constant—the white forsythia (*Abeliophyllum distinchum*) and *Clematis lanuginosa* are mentioned—but in many more cases, a shift in temperatures is needed beginning with three months at low temperatures to spur accompanying low temperature chemical reactions needed before growth can resume. Mr. Deno calls these types of germination "stepwise germinators."

"Needless to say," said Mr. Deno, "it's all a complicated deal with nature holding most of the cards, but with patience we'll be able to properly play out the hand."

TEMPERATURE AND MOISTURE

According to Mr. Deno:

> *A common misconception about seeds is that they will germinate if given a little moisture and warmth. The mistake arises because about 50 percent of the seeds found in the temperate zone can be collected, put in an envelope on a shelf, and germinated months later when given moisture and warmth. What everybody overlooks is that the period of dry storage on the shelf is essential to germination. Why? Because the germination inhibitors are destroyed by the drying process.*

That need for a drying process prevents seed from germinating before dispersal. It is the most common mechanism for destroying germination inhibitors. Most important, it's the dominant process for germination delay in most species of agricultural importance to humans. Early humans were able to put seeds into dry storage over the winter for germination when spring arrived.

Most garden annuals fall into this category, so it was easy for gardeners to think that all seeds can be dry-stored and will germinate immediately upon adding water. The other 50 percent of temperate zone plants were avoided and simply regarded as recalcitrant germinators.

After inhibitors have been destroyed by drying, the seed usually germinates immediately upon meeting moisture at 70°F. For obvious reasons, Mr. Deno includes all these seeds in the D-70 group. Afer one month, *Anemone cylindrica* exhibits only 4 percent germination if sown when fresh, but 98 percent germination if the process follows six months of dry storage at 70°F. *Aster coloradoensis* shows no sign of immediate germination, but a 100 percent rate after six months of being dry. The same holds true for six species of those delightful spring-blooming rock-garden flowers in the genus *Draba: Draba aizoon, D. compacta, D. dedeana, D. densifolia, D. lasiocarpa, D. parnassica,* and *D. sartori* all follow this basic pattern.

There are some seeds that will germinate immediately at 70°F regardless of being sown fresh or following dry storage. *Eunomia oppositifolia*, for example, an evergreen subshrub from the mountains of Asia Minor with charming pink, flowers will follow this course as will *Centaurea maculosa*, a short-lived perennial called the spotted knapweed that is attractive in wild gardens. The question arises about why this seed does not germinate while in the seed capsule, but apparently a tight capsule keeps water away from the seed.

There are seeds that belong to the group called D-40, where the seed requires a period of drying, then germinates at 40°F.

A frequent pattern for seeds is to have an induction period of several weeks followed by a sudden onset of germination. Examples are seeds of the stars-of-Persia (*Allium christophii* [Syn: *A. albopilosum*]) that germinate 85 percent over five to thirteen weeks at 40°F and one of the camass lilies (*Camassia leichtlinii*) that germinates at 100 percent over five to fourteen weeks. The initial induction period of five weeks in both these examples represents the time required for the destruction of the chemical systems blocking germination. They are not timed for diffusion of oxygen and water. The latter usually is complete in a few days and would have led to a gradual onset of germination.

One of the most remarkable effects of temperature and time are seeds that require oscillating temperatures to germinate. Many of the dogwoods (*Cornus* spp.) follow this pattern. The pattern has evolved to foster germination in autumn or spring, when the diurnal temperature changes are the greatest.

Elegant experiments can be constructed. Take *Cleome serrulata*, for example. These seeds will not germinate in the dark if held at 70°F or 40°F, but if the temperatures are shifted back and forth between these at intervals in the range of six to twenty-four hours, the seeds demonstrate 100 percent germination at a rate of 6 percent per day after an induction period of four days.

SEEDS EMBEDDED IN FRUITS

In many cases, nature depends on birds and animals to distribute seed. Nature accomplishes this by embedding seeds in fruits and berries to be eaten directly or as nuts in shells. These can be carried off and either eaten or buried by mammals like chipmunks and squirrels with the digestion processes taking care of any pulpy covering.

The pulp contains germination inhibitors that are carried by a water channel through the seed coat into the embryo where they diffuse. When the remaining pulp is removed, the embryo soon destroys the rest of the inhibitors and germination begins. According to Mr. Deno, when pulp is removed from the red berries of common garden asparagus

(*Asparagus officinalis*), the arils of American bittersweet (*Celastrus scandens*), the syncarps of the mulberry (*Morus alba*), and the berries of some grapes (*Vitis vinifea*) at 70°F, germination begins with a few days. Although, says Deno, it is not an exact science:

> *But the examples of seeds that germinate immediately after washing are the exception. More often than not, seeds from fruits have extended multicycle germinations. For example, I found this pattern in* Cornus, Cotoneaster, Ilex, Prunus, *and* Viburnum. *With these species, the pulp may or may not have germination inhibitors since the seed already has another mechanism for delaying germination until after the seed is dispersed.*
>
> *Some seeds have arils, a fleshy appendage that partially or sometimes completely covers the seed. Examples include many* Iris, Jeffersonia, *and* Trillium. *The aril should be removed as this reduces the chance of mold growing on the towels.*

The aril is designed to attract ants and other insects that carry off the seed with the attached aril, to eat it at their leisure. Two questions arise: Does the aril have to be removed for germination, and do the ants and other insects secrete any chemical that helps germination? After some experimentation, there seemed to be little effect (except for the seed of glory-of-the-snow, where germination only occurred when the arils had been soaked off the seed).

Seed obtained from botanical expeditions, most commercial seed houses, and the various seed exchanges are usually not cleaned. Just as with fresh fruit, the seed must be soaked, washed, and cleaned.

THE PAPER TOWEL

To pursue experiments in which the life processes of literally thousands of seeds were charted, Mr. Deno could not turn to endless peat pots with sterilized growing mediums. He found another way thanks to Margery Edgren, who introduced him to the process of germinating seeds in moist paper towels at an annual meeting of the American Rock Garden Society.

Mr. Deno chose high-strength paper towels that are perforated in approximately one-foot-square sections. These are folded three times (see diagram below) with the species name and description of the experiment written on the outside with an indelible pen for labeling. The towel is wetted until just saturated with water (excess water can easily be pressed out), the seeds are then placed inside, and the folded towel placed within a polyethylene bag (6 1/2-inch by 5 7/8-inch reclosable

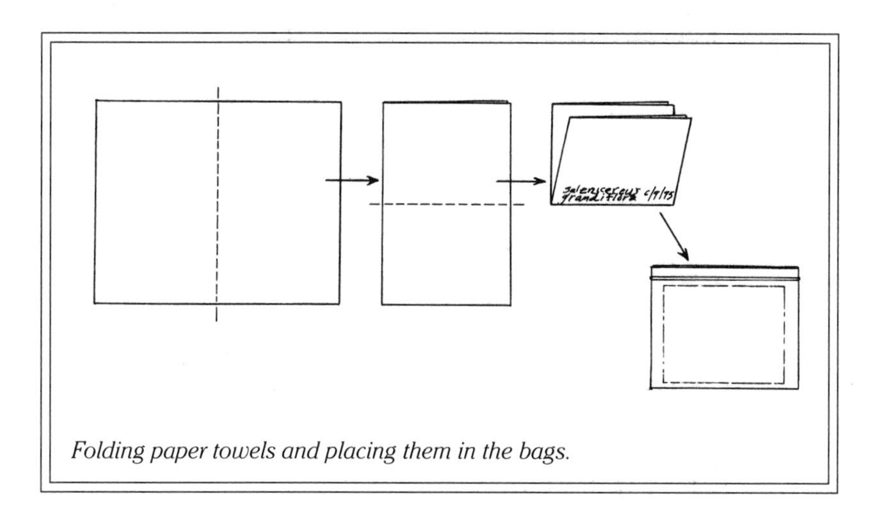

Folding paper towels and placing them in the bags.

sandwich bags are perfect), and the tops are folded over and not sealed. This last instruction is very important. Although polyethylene film can transmit oxygen, it often does not keep up with the amount needed by germinating seeds; the bags should prevent water loss but allow free passage of oxygen.

It was possible to store five hundred of these folded towels per cubic foot—try that with petri dishes or peat pots! The plastic bags were examined as often as each day (always at the same hour) or even at ten-day periods. Additional moisture need only be added at intervals of several months.

Many seeds require light to germinate, and these are often very small seeds whose seedlings cannot tolerate the periodic underwatering and overwatering that almost inevitably occurs in open pots. A highly efficient way to handle this type of seed is to sterilize the surface of the medium by pouring boiling water over it three times. Let it cool, and sow the seeds directly on the surface. Place the pot with its seeds in a polyethylene bag (use the thinnest film like that found in Baggies). Close the top of the bag with a twistem, but do not seal it airtight. Just seal it tightly enough to inhibit water loss. Place the bags under fluorescent lights, remembering that direct sunlight will cook the seedlings. The fluorescent lights should be connected to a simple electric timer that is set for twelve hours of light and twelve hours of dark each day. These bags should be checked every month or so, but they will probably not need additional water for at least six months. At the end of this time the grower has a pot of seedlings with virtually no effort. This technique is also wonderful for other species with tiny seeds such as begonias and gesneriads.

FOLLOWING THE PAPER TOWELS

When the radicles of the seeds emerge, they can be carefully removed from the toweling and planted into pots with each pot full of seedlings, thus saving a great deal of space. According to Mr. Deno, with some species, the towels can be stacked upright and the seedlings allowed to grow on until either the cotyledons or the true leaves emerge.

There are some species that are liable to die at the tips, especially when the temperature shifts from 40°F to 70°F, and according to Mr. Deno, they include doll's eyes (*Actaea pachypoda*) and red baneberry (*A. rubra*); most of the gingers (*Asarum* spp.); glory-of-the-snow (*Chionodoxa luciliae*); winter aconites (*Eranthis hyemalis*); Puschkinias (*Puschkinia scilloides*); trilliums (*Trillium* spp.); and tulips (*Tulipa* spp.).

According to Mr. Deno:

> *There's a possiblity that some chemical in the paper towel is toxic to the seedling, but it seems more likely that because these plants do most of their growing during the cooler part of the year and require temperatures between 40°F and 70°F for the best growth, subjecting them to a sudden shift in temperatures throws the seedlings into a forced dormancy without a storehouse of nutrients and they die. My suggestion with these problem plants is to time the germinations for spring or fall when the seedlings can go outside or sow these seeds in the open and let nature take over.*

THE EFFECTS OF LIGHT

A few seed catalogs mention the need for light for some seeds to germinate, usually listing annuals like coleus (*Coleus* spp.), nicotiana (*Nicotiana* spp.), and bells of Ireland (*Molucella laevis*) and perennials like coralbells (*Heuchera* spp.) and balloon flowers (*Platycodon grandiflorus*). According to Christopher Lloyd and Graham Rice in their book, *Garden Flowers from Seed* (see chapter 13), none of these seeds require light. Turning to Mr. Deno's book, we find that balloon flowers do *not* need light and *Heuchera hallii* (a wildflower from Colorado) required light. None of the annuals were listed. Mr. Deno notes,

> *Photo-effects were far more complicated than described in the available literature. Just finding out that in many species photo-responses are often dependent on the treatment of the seed prior to light exposure made it difficult and led to an expanded series of experiments. One helpful generalization is that photo-effects are largely restricted to swamp and woodland plants because many of these plants grow where there is sufficient moisture so light turns out to be the limiting factor. And in general, photo-effects are not characteristic to families and genera*

> *because many genera contained both photosensitive species and species without any reactions to light.*

Mr. Deno lists six categories. The first is 70L, species that would only germinate at 70°F and in light—these two factors were the only mechanisms needed. Plants included *Astilbe chinensis*, the swamp hyacinth (*Helonia bullata*), and both the cardinal flower (*Lobelia cardinalis*) and the great blue lobelia (*L. syphilitica*). Oconee-bells (*Shortia galacifolia*) also germinated only in light, but germination was complicated by an intolerance of dry storage, whereas all the old wildflower books say that only fresh seed will work.

The second catagory is D-70L and 40-70L. In both groups, germination occurs only in light, but the seeds require a pretreatment before they will germinate in 70L. In the D-70L group, six months of dry storage at 70°F is needed. In the 40-70L group, three months at 40°F was needed first. D-70L germinators include sweetbells (*Leucothoe racemosa*), whereas 40-70L germinators were swamp milkweed (*Asclepias incarnata*) and blue vervain (*Verbena hastata*).

The third category is 70D and consists of species whose germination was totally blocked by light. One species of interest is *Cyclamen persicum*. I have a twenty-one-year-old plant in my greenhouse that never has fewer than 150 blossoms, and many form seed after flowering. Last year the pods opened and spilled seed on the greenhouse floor that consists of at least a three-foot layer of inch-long pieces of gravel. With the warming of spring, I noticed dozens of little plants, but their leaves were on long white petioles coming from far down in the rock. None of the visible seeds had sprouted.

The fourth category lists seeds that must have light for germination if the seed is fresh and the temperature is 70°F. But after dry storage, many of the light requirements were not needed.

The fifth category is 70L(O) and consists of species that germinate in either 70°F and light or in outdoor conditions. According to Mr. Deno, germination fails in all other treatments.

The sixth category is 70(L), where light promoted germination but was not absolutely required. It includes plants like *Clematis connata* from the Himalayas, the Eastern sycamore (*Platanus occidentalis*), and the Catawba rhododendron (*Rhododendron cawtawbiense*).

STORING SEEDS

Both gardeners and scientists have been interested in the life span of stored seeds for hundreds of years. The public's imagination is often sparked by reports of seeds of plants like the regal lotus, which have been in tomb storage for three thousand years, only to spring to life in a late-twentieth-century laboratory. Or seeds that have spent decades in old herbarium cupboards and have sprouted when given water. But according to Mr. Deno, dry storage is ultimately fatal to all seeds; it's only a matter of time:

> *A few species have seed that remains viable for only a few days with most seeds retaining vitality for a year or two. Some species obviously have seeds that last for ten to fifty years, but usually the viability declines, and although the seed may germinate, the germination rate is low, and often the seedlings are either abnormal, lack vigor, or both.*
>
> *But in my experiments, many seeds were killed by dry storage at either 40°F or 70°F. They include the rue anemone* (Anemonella thalictroides), *spring-beauty* (Claytonia virginica), *the beautiful and fragile* Jeffersonia diphylla, *and the Welsh poppy* (Meconopsis cambrica).
>
> *At this stage in my work, I would suggest that for the short term, seeds be stored in the refrigerator at 40°F because death rates are generally about a third as fast at 40°F as at 70°F. For seeds that germinate at 70°F but have a zero rate of germination at 40°F, the seeds can be stored in a moist paper towel within a polyethylene bag in the refrigerator at 40°F. Just fold the bag over several times and never seal it tightly.*
>
> *A beautiful example is the arctic willow* (Salix arctica). *Seeds are completely dead after just three weeks of dry storage at 70°F, but they can be stored moist at 40°F for at least three months. On shifting to 70°F, they germinate just as if the three months at 40°F had never occured.*
>
> *Conversely there are species that can be stored moist at 70°F. An example is the winter aconite* (Eranthis hyemalis). *The seeds actually require three months moist storage at 70°F before they will germinate at 40°F, but the period at 70°F can be extended up to twelve months. The seeds will still germinate at 40°F.*

SEED COLLECTING

Not all seeds are as easy to collect as opium poppies (*Papaver somniferum*), where the seed capsule acts like a saltshaker, or columbines (*Aquilegia* spp.), where you simply turn the seed capsule over on a white paper, and the seeds pour out. If milkweeds (*Asclepias* spp.) are not collected at the proper time—just before the seed pod splits—the slightest breath of wind will send the silky parachutes across the backyard and beyond your reach.

"For example," said Mr. Deno,

> *the seeds of penstemons* (Penstemon *spp.) are enclosed in a hard capsule with a sharp beak. You can crush the capsule with pliers after which the seeds will shake out. Then to separate the seeds from the capsule debris, shake the mixture in a cupped sheet of smooth paper. The debris will rise*

to the top and the seed to the bottom. Then lay the sheet flat down and brush the debris to one side. This procedure works with all kinds of seed capsules. Seeds in fruits can be washed.

CONCLUSIONS

When a gardener is confronted with the wealth of material gathered by Mr. Deno, he or she can have two reactions, the first to go on as usual, sowing seeds the same old way. It's like treating disease under the Doctrine of Signatures; you trust to luck and by the law of averages a few will succeed, a few will fail, and the rest will muddle through.

But there's a second course if the gardener is willing to strike off in a new direction (keeping careful notes as the days pass). I, for one, am indebted to Mr. Deno, not only for his scholarship, but also for his popularizing the concept of planting in paper towels. No longer is sowing seed the chore it once was, and a drawer or box can hold what once took up entire bench tops in an already crowded greenhouse.

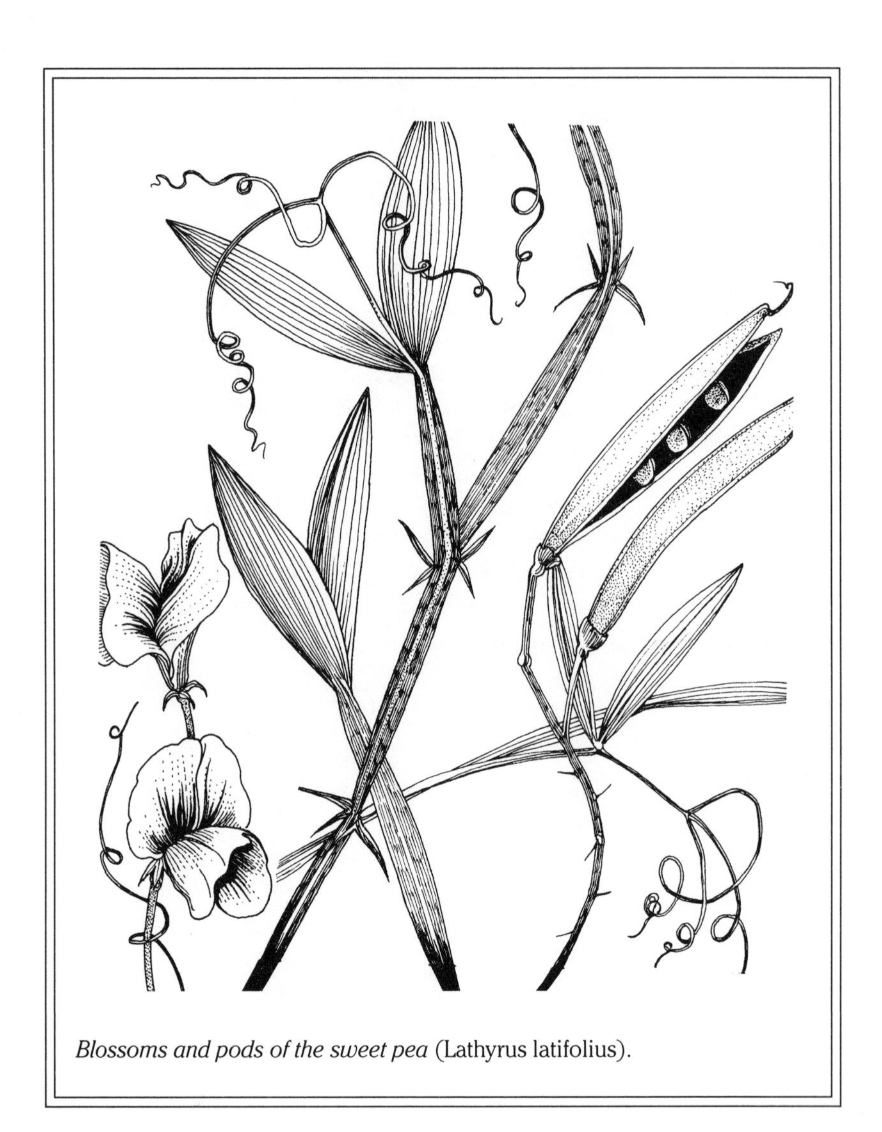

Blossoms and pods of the sweet pea (Lathyrus latifolius).

CHAPTER 9

The Buying and Selling of Seeds

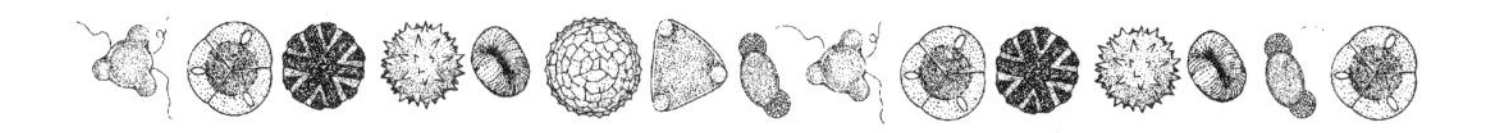

The genus *Adonis* takes its name from a god of Greek mythology. The beloved of both Venus and Persephone, Adonis was a god of great importance in many religions, including the worshipers of Phoenicea and in Babylonia, where his name was Tammuz. His mother, Myrrha, was banished for an incestuous relationship with her father, Cinyras, an Assyrian king, and forced to wander the earth alone. In the midst of giving birth to Adonis, she was changed into a myrrh tree (*Balsamea myrrha*).

Adonis was too handsome for his own good, and many of the Greek hierarchy were jealous in one way or another. After all, the machinations of the Olympic gods were the stuff of yesterday's legends—and today's most widely read novels.

One afternoon while hunting, Adonis was confronted by a particularly menacing wild boar and was pierced by the boar's tusks. Venus, flying high overhead in a chariot led by a team of flying cygnets, looked down to see the blood pour onto the ground. She reasoned that if Adonis's mother could be turned into a tree, he could be remembered in much the same way. So she sprinkled the blood with nectar's dew, and one hour later a blossom grew, "born from the blood, itself of sanguine hue." But like Adonis, the blossoms are short-lived: Even before the wind can blow the blossoms open and scatter the petals, *Adonis flammeus*, growing wild in the wheat fields of Europe, is cut down by advancing mowers.

Adonis was so loved by the women of Greece, and the gods were so disheartened by the sorrows of Venus, that he was required to spend only a third of the year in Hades, leaving him free to roam the upper world the rest of the time. So when midsummer arrived, festivals of Adonis sprang up, and the first gardens of Adonis appeared. These gardens consisted of baskets or pots of terra-cotta in which the quick-growing plants were sown, tended for eight days, allowed to wither, and then flung into the sea—or into local springs or pools—along with figurines of the dead youth. So thousands of years ago, seeds were sold for aesthetic pleasures.

Columbus brought seeds from the Old World on his second voyage in 1493, carrying wheat, barley, sugarcane, and grapes. And the colonists that followed the Spanish armies into Florida, Mexico, and Peru took along seeds of their favorite crops. Because they tended to settle in areas that reminded them of home, most of these plants adjusted with ease.

But as the population grew, more seeds were needed. The Spanish government ordered that all ships sailing for the Indies carry both plants and seeds in their cargos. As a result, almost 150 species and varieties were introduced to New Spain, including lemons, cabbage, turnips, anise, alfalfa, flax, bamboo, daffodils, iris, poppies, and of all things, carnations. The Indians even helped in the distribution of plants by carrying seeds on their travels. The wild peaches that were found by the first settlers of Pennsylvania probably came from the original trees planted a century earlier in St. Augustine, Florida.

Of course, the trading worked both ways, and the Spanish took back Indian crops including corn, tobacco, cotton, chili peppers, squash, and tomatoes.

Then in 1611, Jamestown planters brought tobacco from Trinidad. So by the end of the seventeenth century, most of the food and feed crops growing in present-day America were already established.

At that time it was believed that America should be growing cash crops for Europe, and the choices included rice, indigo, cotton, sugar, spices, tea, and mulberry trees for silkworms. However, few of the plants introduced from the subtropics could withstand the winters found in most of the settled country.

Then in the 1690s, South Carolina planters brought seed for rice from Madagascar, and it proved a great success growing on their lowlands around Charleston. And in 1745, Eliza Lucas brought indigo seeds from Antiqua, West Indies, and they, too, proved successful.

Just after the end of the American Revolution, seeds of sea-island cotton were introduced from the Bahamas, adding a fourth cash crop to tobacco, rice, and indigo. Unlike regular cotton, the lint of sea-island cotton separated easily from the seed (although this was no longer necessary after Eli Whitney invented the cotton gin in 1793).

On 14 May 1686, a gardener hired by William Penn wrote to England about gardening in America, and he commented, "Trees and

Bulbes are shot in five weeks time, some one Inch, some two, three, four, five, six, seven, yea some are eleven Inches . . . And seeds do come on apace; for those seeds that in England take fourteen days to rise, are up here in six or seven days." The letter concludes with a request: "Pray make agreement with the Bishop of London's Gardiner or any other that will furnish us with Trees, Shrubs, Flowers and Seeds, and we will furnish them from these places."

Unlike most American cities, Penn had great plans for mapping gardens of beauty in his state, including in the city of Philadelphia. There were to be five public squares inside the town proper with a great square of ten acres in the center. And he directed that every man's house "be placed, if the Person pleases, in ye middle of its place as to the breadth way of it, so that there may be grounds, on each side, for Gardens, or Orchards, or fields, yet may be a greene Country Towne, wich will never be burnt and will always be welcome."

But Philadelphia aside, according to Ann Leighton in *American Gardens in the Eighteenth Century* (Houghton Mifflin, 1976), in the American colonies, "the seventeenth century gardens had been almost totally what it is now fashionable to call 'relevant.' They existed to feed, clothe, clean, cure, and comfort the settlers."

America was on the march, and by the eighteenth century things began to change, reflecting what was happening in England and Europe. Both seeds and plants were on the market. One of the first seed shops in Paris was opened by Pierre Janfroid in 1728, and his successor Andrieux issued his first catalog in 1769. Ten years later the Vilmorins married into the family, and since 1799, the firm of Vilmorin-Vilmorin-Andrieus et Cie has been selling seeds around the world. Germany saw the formation of the seed house of L. Spath in 1720, continuing well into the twentieth century.

Many of the seeds imported during the eighteenth century were grown in the gardens of renowned plantsmen like John Bartram of Philadelphia. Bartram's garden was the most widely known of these seed plantations, and he specialized in plants of this continent. While traveling abroad, men like Benjamin Franklin and Thomas Jefferson sent seeds to Bartram. Franklin, for example, introduced two Scottish crops, rhubarb and kale. Jefferson risked the death penalty in northern Italy by smuggling seeds of an upland cotton that the Italians were guarding for a monopoly.

Although seeds were traditionally bought and sold for agricultural purposes, it was not until the rising of the merchant class of the last few hundred years that seeds for ornamental purposes were available in shops.

The following advertisement appeared in the 9 July 1763 issue of the *South Carolina Gazette*:

Thomas Young, Sen. has at present a fine assortment of kitchen garden seeds, several kinds of grass seeds, flower seeds and roots, flower glasses, and some garden tools., Lists of which may be seen at his house near the west end of Broad Street. He will sell no seeds but such sorts as have grown in his garden, and therefore can be warranted good. Those gentlemen or ladies who want flower roots, must call for them soon as most of them are now out of the ground.

In her book *Old Time Gardens* (Macmillan, 1901), Alice Morse Earle wrote a great deal about the early sales of seeds in America.

> *The shrewd and capable women of the colonies who entered so freely and successfully into business ventures found the selling of flower seeds a congenial occupation, and often added it to the pursuit of other callings. I think it must have been very pleasant to buy packages of flower seed at the same time and place where you bought your best bonnet, and have all sent home in a bandbox together; each would prove a memorial of the other; and long after the glory of the bonnet had departed, and the bonnet itself was ashes, the thriving Sweet Peas and Larkspur would recall its becoming charms.*

An ad in the *Boston Gazette* of February 1719 read:

> *Garden Seeds. Fresh Garden Seeds of all Sorts, lately imported from London, to be sold by Evan Davies, Gardener, at his house over against the Powder House in Boston; As also English Sparrow-grass Roots, Carnation Layers, Dutch Gooseberry, and Current bushes.*

A seedswoman advertised the following list of seeds in a Boston newspaper on 30 March 1760:

> *Lavender, Palma Christi, Cerinthe or honeywort, loved of bees, Tricolor, Indian pink, Scarlet Cacalia, Yellow Sultans, Lemon African Marigold, Sensitive Plants, White Lupine, Love Lies Bleeding, Patagonian Cucumber, Lobelia, Catchfly, Wing-peas, Convolvulus, Strawberry Spinage, Branching Larkspur, White Chrysanthemum, Nigaella Romano, Rose Campion, Snap Dragon, Nolana prostrata, Summer Savory, Hyssop, Red Hawkweed, Red and White Lavater, Scarlet Lupine, Large blue Lupine, Snuff flower, Caterpillars [see page 100], Cape Marigold, Rose Lupine, Sweet Peas, Venus' Navelwort, Yellow Chrysanthemum, Cyanus minor, Tall Hollyhock, French Marigold, Carnation Poppy, Glove Amaranthus, Yellow Lupine, Indian Branching Cox-combs, Iceplants, Thyme, Sweet Marjoram, Tree Mallows, Everlasting Greek Valerian, Tree Primrose, Canterbury Bells, Purple Stock, Sweet Scabiouse, Columbine, Pleasant-eyed Pink, Dwarf Mountain Pink, Sweet Rocket, Horn Poppy, French Honeysuckle, Bloody Wallflower, Sweet William, Honesty to be sold in small parcels that every one may have a little, Persicaria, Polyanthos,*

Fifty different sorts of mixed Tulip Roots, Ranunculus, Gladiolus, Starry Scabiouse, Curled Mallows, Painted Lady topknot peas, Colchicum, Persian Iris, Star Bethlehem

In the *Boston News-Letter* of 5 April 1764, another seedsperson, Anna Johnson, advertised a sale at her shop at the Head of Black Harre Lane, leading up from Charlestown Ferry. The list included the following:

A fresh assortment of Garden Seeds, Peas and Beans, among which are, early charlton, early hotspur, golden hotspur, large and small dwarf, large marrowfat, white rouncevals rose and crown, crooked sugar, and grey Pease; large Windsor, early hotspur, early Lisbon, early yellow, six-weeks long podded, and white Kidney Beans; early Dutch, yorkshire, sugar-loaf, battersea, savory, and large winter Cabbage; early and late Colliflower, early orange, scarlet and purple carrots; best smelling Parsnips; Endive, Cellery; Asparagus, and Pepper; early prickly, long and short cluster, white and green turkey Cucumber; Thyme and Sweet Marjoram; Balm, Hyssop and Sage; London short and Salmon Raddish; Lavender; green and white goss, green and white silesia, imperial, cabbage, tennis-ball, marble, and brown dutch lettice; ripe canary Seeds; red and white Clover; herd's Grass; red top and tye grass Seeds; also a Parcel of curious Flower-Seeds.

That list is amazing when one remembers that outside of the wealthier inhabitants of the cities, most Americans were working round the clock, 350 days a year, taking time out only for church, Easter, and Christmas.

The first seed house in America was opened by David Landreth & Son of Philadelphia in 1780. At first, the seed industry grew with little exuberance, and by 1850 there were forty-five seed firms, mostly in the East. Seeds for many of the vegetables grown in this country before the Civil War were brought over by immigrant wives who saved seed from each year's harvest.

In 1817, Elkanah Watson, one of the founders of the New York Society for the Promotion of Useful Arts, sent a series of letters to all the American counsels in Europe, asking for seeds. In answer to one request, the counsel in Valencia, Spain, sent seeds of fourteen types of wheat, one of barley, and one of oats.

Kohlrabi and the rutabaga were introduced to American tables at the end of the eighteenth century. And with the rise of the merchant seed houses, competition began in earnest. In 1950, U. P. Hedrick wrote in his book, *A History of Horticulture in America to 1860* (Timber Press, 1988), about the following incident in New York: "A curious dispute arose in New York papers in 1818 between Grant Thorburn, an early seedsman, and William Cobbett, an owner of a New York seed store in 1818. Cobbett claimed that he introduced rutabaga and sold better seed at a dollar a pound than Thorburn; but Thorburn retorted that his seed was better at the same price, arguing that in 1796 a large field of these turnips was grown by Wm. Prout on what was then occupied by the navy yard, at the city of Washington." Thorburn was probably right, but in Cobbett's *American Gardener* and *A Year's Residence in the United States of America*, he was a great champion of the rutabaga.

Because of disputes, it was time for the government to get involved. The notice sent to American counsels asking for seeds was made official in the Congress of 1825, authorizing the counsels to send home seeds of rare plants. In 1836, the United States Patent Office was started under Henry Leavett Ellsworth as director. According to accounts, he was more interested in agriculture than inventions, collecting rare and valuable varieties of seed the first week he assumed office. The seeds were then distributed with 40,000 packages going out in 1840 and 2,333,474 packets in 1877, with some of them being gifts of European governments.

U. P. Hendrick wrote that "the distribution of seeds became more and more a species of petty graft on the part of congressmen who used their franking privileges to send them to those who would take them. Instead of rare varieties, the packages often contained packets of the commonest and cheapest seeds that conniving seedsmen could supply."

Billions of seed packets were distributed, mostly containing seeds of vegetables and flowers, but they also included seeds of sorghums, sugarbeets, soybeans, and many more. Complaints about poor germination rates persisted and led to questions about the reliability of supply. Some horticulturists questioned the wisdom of selecting seeds for sugar and tea as the most urgently needed crops.

"It was not until well into the twentieth century," continued Hendrick, "that, under the combined forces of farmers, gardeners, seedsmen, and farm and horticultural papers, free seed distribution with its many scandals was brought to a close."

Gardeners at the end of the twentieth century might wonder about the lure of seeds to the voters, but back then the farmer ruled supreme. And if Congressman Potatohead of Musslebend, Missouri, thought enough to send seeds to the bank president's wife, he really must be quite a guy. After all, crop seeds were indispensable to nine out of ten households in the early days of the country. Most of the seed was homegrown, and seeds for new land were obtained by buying from seed houses or in trade from other farmers.

Patricia M. Tice in her museum catalog, *Gardening in America* (The Strong Museum, 1984), mentions the cost of seeds measured against the cost of plants. She writes:

A wider interest in ornamental gardening developed as leisure time, inexpensive gardening products, and seeds became available. The drastic difference between prices of nursery plants and mail-order seeds can be seen in a price comparison between the Ellwanger and Barry Mt. Hope Nurseries and the Ferry Seed Company of Detroit, Michigan.

Table 1. Price Comparison between Plants and Seeds

	1840, price per plant	***1886, price per seed pack***
Dahlia	$ .50	$.15
Chrysanthemum	1.50	.05
Carnation	.75	.15
Honeysuckle	.50	.05
Hollyhocks	.12	.05
Oleander	1.00	.10
Peony	1.00	.25
Primrose	.50	.25
Wallflower	.75	.10

Joseph Breck and Sons (then Company) began operations in 1837, and in 1851, Breck published *The Flower Garden*, or *Breck's Book of Flowers*, a book about seeds that became so popular it ran through five editions. At the same time, Fearing Burr (1815–97), a well-known seedsman and writer, published *The Field and Garden Vegetables of America* (reissued at Chillicothe, Illinois: *The American Botanist,* 1988), a book that contained full descriptions of nearly eleven hundred species and varieties of plants and that included directions for propagation and the uses of vegetables. This book is a treasure-trove of information about eighteenth-century farms and gardens, especially because it documents so many cultivars that have now disappeared from commerce (see page 99).

In 1847, the Landreth Company of Philadelphia bought 375 acres outside of Bristol, Pennsylvania, hiring over 125 employees to run its seed operations. They needed a separate business just to provide paper bags and envelopes. By midcentury, Landreth's had a worldwide reputation, and their seeds were planted in every state of the Union, arguably, wherever gardens were maintained. David Landreth was one of the founders of the Pennsylvania Horticultural Society, an organization that still inhabits Walnut Street in Philadelphia and maintains a worldwide reputation.

In 1853, when Commodore Perry lifted the Bamboo Curtain that surrounded Japan, the Landreth Company sent along a box of seeds. In return, they received a box of Japanese seeds, a gesture that marked the beginning of seed trade with Asia. Over the years, trading revealed so many valuable plants that scores of pages would be needed for its documentation.

However, 1853 turned out to be a year of change in the seed business for another reason. B. K. Bliss, a native of central New York, moved to Springfield, Massachusetts, and opened the first mail-order seed business in America. Later, the Bliss Company was also the first to print a catalog with colored plates, and over the following years, it brought many new garden plants to the American gardener (although some argue that James Vick, a Rochester seedsman and innovator, might have been the first when his 1864 catalog featured a colored lithograph of double zinnias that had been developed only four years earlier). In 1858, Bliss described 676 varieties of flower seeds alone. Seven years later the list had grown to 1,612.

Other nurseries, too, began using color in their catalogs. It became such a popular move that in 1886, when the Ferry Seed Company published its new catalog without colored illustrations, the company felt responsible in printing the following statement on the first page:

> *[I]f it seems lacking in brilliantly colored plates of impossible vegetables and glowing descriptions of the surperlative excellence of new sorts which are meant to revolutionize garden practice, it is because our aim has been to give in its pages information which will enable our readers to have a good garden, rather than to tempt them to purchase at exorbitant prices a few seeds of some untried novelty liable at least to result in failure and disappointment.*

England's Thompson & Morgan Company has been producing consecutive seed catalogs since 1855, equalling 145 years of service.

The interest in new seeds and plants was a bottomless pit. In 1872, Peter Henderson & Co. of New York City began publishing two catalogs; one for general horticultural needs and one dedicated to new plants of that particular year (see page 96).

Speaking of James Vick, Miss Tice mentions a book published in 1877, titled *History of Monroe County*, that gave an intimate look into the large scale of his Rochester nursery:

> *In addition to the ordinary conveniences of a well regulated seedhouse, there is connected with this establishment a printing office, binder, box-making establishment, and artists' and engravers' rooms, everything but the paper being made in the establishment. The machinery necessary for the various departments is driven by steam-power . . . [and] the magnitude of this institution is illustrated from the fact that it occupies a building five stories in height, including a basement 60 feet in width*

and 150 feet in length, with an addition in the upper story of a large room over an entire adjoining block.

The first floor is used exclusively as a store for the sale of seeds, flowers, plants, and all garden requisites and adornments, such as baskets, vases, lawn-mowers, lawn-tents, aquariums, seats, etc., etc.

The second floor is devoted to business offices . . . [and] the mail-room is upon this floor, and the opening of letters occupies the time of two persons, and they perform the work with astonishing rapidity, often opening three thousand in a single day. After these letters are opened they are passed into what is called the registering room, on the same floor, where they are divided into States, and the name of the person ordering and the date of the receipt of the order registered. They are then ready to be filled, and are passed into a large room, called the order room, where over seventy-five hands are employed, divided into gangs, each set or gang to a State, half a dozen or more being employed on each of the larger States. After the orders are filled, packed, and directed, they are sent to what is known as the post-office . . . where the packages are weighed, the necessary stamps put on them, and the stamps cancelled, when they are packed in post-office bags, furnished by the government, properly labeled for the different routes, and sent to the postal cars. Tons of seeds are thus dispatched every day during the busy season.

On the third floor is the German department, where all orders written in the German language are filled by German clerks; a catalog in that language is also published. On this floor, also, all seeds are packed, that is, weighed and measured and placed in paper bags, and stored ready for sale. About fifty persons are employed in this room, surrounded by thousands of nicely-labeled drawers.

On the fourth floor are rooms for artists and engravers, several of whom are kept constantly employed in designing and engraving for catalogues and chromos [chromos were colored reproductions that used lithographic stones]. Here, adjoining is the printing office, where the catalogue is prepared and other printing done, and also the bindery, often employing forty or fifty hands, and turning out more than ten thousand catalogues in a day. Here is in use the most improved machinery for covering, trimming, etc., propelled by steam.

The immense amount of business done may be understood by a few facts: Nearly one hundred acres are employed, near the city, in growing flower seeds, mainly while large importations are made from Germany, France, Holland, Australia, and Japan. Over three thousand reams of printing paper are used each year for catalogues, weighing two hundred thousand pounds, and the simple postage for sending these catalogues by mail is thirteen thousand dollars. Millions of bags and boxes are also manufactured in the establishment, requiring hundreds of reams of paper and scores of tons of pasteboard.

Miss Tice then went on to point out that the Shakers at Mount Lebanon in Pennsylvania are generally given credit for introducing small packets or "papers" of seed, beginning this selling method during the first quarter of the nineteenth century.

Charles Van Ravenswaay, in his book, *A Nineteenth-Century Garden* (Universe Books, 1977), notes that not all the progress was made on the East Coast.

> *West Coast plant needs and interests were stressed in the assortment of seeds, plants and bulbs sold by E. E. Moore in San Francisco in 1871, which included all those of "known excellence, as adapted to the various localities of this country and adjacent States." Most of the familiar varieties were represented, along with forty-two varieties of sweet, pot, and medicinal herb seeds. He also sold opium poppy seed (to be grown for medicinal purposes) and seeds of native California plants.*

During the early part of the twentieth century, the seed industry moved along. Discoveries continued in the field of crop plants and vegetables. New varieties of rice were introduced for the southern farm, and Frank N. Meyer, a young Dutchman, walked thousands of miles in Asia between 1905 and 1918, bringing back zoysia grass for lawns of the burgeoning middle classes.

I'm lucky to have two wonderful seed catalogs in my library. The first is a *Catalogue of Seeds for 1932*, published by the Deposit Seed Company of Deposit, New York. The second is *Everything for the Garden*, published by Peter Henderson & Co., founded in 1872, located in Manhattan on Cortlandt Street. Between these two catalogs, a gardener can learn most of what was happening in American gardens during the Great Depression.

The cover for the Deposit Seed company is in full color and depicts ten plants for a Good Perennial Garden costing only fifty cents, including postage. The choices are Gold Medal Delphinums, Gold Medal Strain Hollyhocks in a wide variety of colors, including crimson, scarlet, white, orange, salmon, pink, lemon, and more; Digitalis, from the curious 'Monstrosa' to the 'Giant Shirleys' to the delicately flowered 'Gloxinia Flowered'; Mixed Lupines; Gold Medal Mixed Columbine, including long-spurred hybrids from Holland and a large choice of both double and single sorts; Pyrethrym both single and double; Giant Daisies or Bellis; Campanula Carpatica or the Carpathian Harebell; Orange Scarlet Oriental Poppy with blossoms from two to four inches across; and Coreopsis, with single, double, and doubles with variously colored centers.

A packet of the big and blousy Roggli Swiss Giant Pansy is described as being huge, with rich colors of sulphur, golden yellow, carmine, orange, scarlet, blue, and white. From the photo of 'Superbimissa', they look as current as anthing offered today. And I wouldn't mind trying the California poppy offered as the 'Geisha' cultivar colored with a

Deposit Seed Company catalog.

rosy orange red on the outside and a rich golden inside. There are more sweet peas offered in this American catalog than the current Thompson & Morgan catalog from England, where sweet peas are both king and queen.

I also wouldn't mind trying the crested cosmos or above all else, a pink form of the always-popular moonflower (*Ipomoea alba*), one that I've never seen offered in twenty-five years of gardening.

As for Henderson's catalog, it numbers 176 pages with full color throughout and more than 60 pages of flower seeds. In 1932, the first year of the All-American Selection of Flower Seed Novelties, nine varieties were offered for two dollars. *Mimulus* 'Red Emperor' sown in March, for example, will provide bloom in late June, and the Sunshine Asters featured have magnificent yellow centers surrounded by petals of blue, pink, white, crimson, violet, or mixed colors. They also offer balloon vines (*Cardiospermum halicacabum*) that today are found only at specialty nurseries.

Peter Henderson & Co. catalog.

Using the California poppies as an example, Henderson's lists nine different cultivars, starting with 'Crimson King', a flower with bright crimson petals and a satiny carmine interior. Also listed is 'Dainty Queen', with flowers of creamy blush tinted coral pink, deepening toward the edges. Or how about 'Toreador Double' with the flower center of golden bronze and the reverse being orange crimson. They are all floral jewels that are missing from the majority of the contemporary big-time catalogs.

It's 1995 as I write this chapter. I have dozens of fancy seed catalogs layered on my desk, and don't get me wrong, I'm delighted to have them. But when I compare today's offerings with those of yesteryear, I begin to wonder whether the modern world is everything it's cracked up to be.

CHAPTER 10

A Hedge Against Starvation: Storing Seeds for the Future

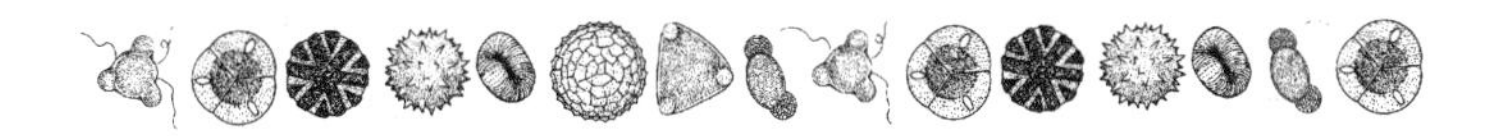

It was a hot day in late July, and a number of us, all potential docents, were following tour guide Anne Tansey through the Botanical Gardens at Asheville, North Carolina. Suddenly, she stopped beside a propagation bed and pointed to a tall tomato vine hung low with round and firm fruits, each the size of a Ping-Pong ball.

"These originally came from my grandmother's garden," she said, "where they grew for more than forty years. I've never found them offered in any of the seed catalogs. Just try them."

She was right; they were uncommonly juicy and sweet, a delightful treat when even the deep shade of the woods felt like a closed car left in the summer sun. We were all promised seeds of that special tomato, and now as a result, that particular strain should be in no danger of disappearing, at least for the next few years.

But many other farm and garden seed varieties—as well as those of us who might have eaten them—have not been so fortunate. More than 90 percent of the vegetable and fruit varieties available at the turn of the century now appear to be extinct.

One hundred and twenty-five years ago, according to Fearing Burr's *Field and Garden Vegetables of America*, 151 varieties of peas were being grown, 82 different turnips, an astonishing 86 types of broccoli, and 112 types of lettuce. There was not only a surfeit of sorrel, but also more than seventeen kinds of endive (*Cichorium endivia*) available for the garden, including the large Batavian, the Dutch green curled, and the staghorn endive.

This year, one major seed catalog listed only six broccolis and twenty lettuces. Although some small seed companies still carry a much larger number of cultivars, the majority of American gardeners aren't exposed to the more diversified lists.

Take peas, for example. The 'Early Warwick', described in Burr's book as being the "head of early peas," was at that time considered to be identical with 'Early Frame', and "for a long period, the most popular of all early varieties . . . [is] at present less extensively cultivated, having been superseded by much earlier and equally hardy and prolific sorts." So one wonders about those many early prolific peas and just how tough they were in surviving disease and drought.

Even in 1865, Burr noted that some popular and hardy varieties were disappearing from the market, continuing the unfortunate trend toward fashionable plants. Every year, new cultivars are brought in to replace the old, following the belief that gardening consumers, just like automobile dealers, must have a new model every year. Every species, variety, and cultivar that disappears is a loss of a link with the past—a link that is not only important in a cultural and aesthetic sense, but also in a biological one. It means a loss of germ plasm, or genetic material, and regardless of dreams of *Jurassic Park*, it can never be replaced.

That seeds have been important to the growth of America, there is no doubt. In the 1600s, books written by New England's John Josselyn contained careful lists of the vegetables that he found in profusion. And according to Alice Morse Earle in her charming book *Old Time Gardens*, fifty years after the Pilgrims landed "the country was well stocked with vegetables," and every ship that arrived from Europe brought more seed.

For example, at the beginning of the last century, there were eleven different types of French sorrel (*Rumex* spp.), numbering three species and eight cultivars, found in American gardens.

And if that were not enough, not only have hundreds of different kinds of vegetable and salad seeds disappeared (not to mention fruits and flowers), but also America's sense of humor—at least at the table—also seems to be on the decline.

Burr's book describes three pseudo-insects meant to liven up Civil War salads, including caterpillars, snails, and worms.

The caterpillar plant (*Scorpiurus vermiculata*) is a hardy annual and a member of the Legume family. The flowers are yellow, streaked with red, and not too attractive, but the seeds are produced in green pods that strikingly resemble their namesakes. In addition to the common caterpillar, there were three other species: the furrowed caterpillar, with brown pods and gray green furrows; the prickly caterpillar with one-quarter-inch pods of brownish red with shades of green; and the villous or hairy caterpillar, larger than the prickly but marked with small points placed along longitudinal ridges.

The caterpillar plant from Gaerarde's Herbal, *1597.*

"No part of the plant is eatable," wrote Burr, "but the pods, in their green state, are placed upon dishes of salads, where they so nearly resemble certain species of caterpillars as to completely deceive the uninitiated or inexperienced."

Snails were *Medicago orbicularis*, another annual with seed pods that distinctly resembled some species of snails to a remarkable degree and "are placed on dishes of salad for . . . pleasantly surprising the guests."

Worms (*Astragalus hamosus*) are another annual plant with pods that resemble, in their green state, some descriptions of these animals.

Just as in the animal world, the heredity of plants is controlled by the genes and chromosomes found in the nucleus of the cell. It is these encoded combinations of amino acids that enable one tomato to develop firm flesh that resists unnecessary bruising when shuffled about during shipment to the marketplace. Another strain may barely withstand being moved from the backyard to the dinner plate, and yet another turns out to be sweet and juicy like Anne Tansey's grandmother's.

Even more important in ensuring food quantity are genes that may make wheat resistant to drought or corn resistant to fungus. The decreasing diversity among the world's agricultural plants has frightening implications. Don't forget that one-fourth of the human energy demands of the United States and three-quarters of the human diet of the rest of the world are provided by cereal seeds. We also consume vast numbers of unprocessed seeds such as pecans, peanuts, and pistachios. We mash, grind, boil, and compress other seeds into margarine, medicines, cosmetics, and alcoholic beverages.

The seeds of wheat provide more food for the human race than any other plant or animal. Wheat originated in Mesopotamia and the Mediterranean. For centuries, its germ plasm moved through generations of farmers into new environments, being altered not by human manipulation but by the principle of survival of the fittest. These primitive cultivars, called landraces, possessed great advantages for their limited ranges. Wheat grown in dry climates tended to become drought tolerant. Wheat that could be successfully grown where certain pests prevailed was naturally tolerant to those pests.

But around 1930, the genetic sciences began to come of age. Breeders developed hybrids for higher yields and made them widely available. Even as the amount farmland decreased, production soared, and old varieties, with the disease and drought resistances they had built over centuries, began to disappear.

What can happen when natural diversity disappears is illustrated by the Irish potato famine of the 1840s. Nearly two million of Ireland's poor starved to death because none of the handful of potato varieties on which they depended for sustenance were resistant to a Mexican potato blight, *Phytophtora infestans.*

The United States' most famous crop disaster was less tragic, but it did sound a warning. In 1970, a fungus, the Southern corn leaf blight (*Bipolaris maydis*), was able to wipe out 15 percent of the nation's corn crop because nearly all the varieties being grown had come from the same parent. The blight was quickly brought under control by replanting corn fields with older hybrids that remained in seed company reserves.

The presence of these seeds in commercial labs was fortunate. In the United States, very few crop seeds are ours for the collecting. Not a single major food or fiber plant grown in this country today originated here. Wheat arrived via Europe and southwest Asia, rice from southeast Asia, and corn from Central America. About the only native plants now used for food in any significant way are sunflowers, Jerusalem artichokes (*Helianthus tuberosus*), blueberries, cranberries, some nuts, and a few beans and squashes.

Even in their countries of origin, the wild relatives of our crops are disappearing. Destruction of habitats such as rain forests and fields and the replacement of indigenous crops by American hybrids is causing valuable genetic material to disappear at an alarming rate.

The same scientists who were creating the hybrids that drove native varieties to extinction finally awakened to the problem of lost germ plasm: They saw that they would soon breed their products into an evolutionary corner. In 1944, the National Research Council recommended that the U.S. Department of Agriculture establish what would essentially be a bank to preserve plant genetic material in case it should be needed in the future. Two years later, Congress passed the Research and Marketing Act, giving the federal government responsibility for obtaining foreign seeds and plants, while the states' duty was to evaluate them. In 1947 and 1948, four regional plant introduction stations opened in Ames, Iowa; Pulman, Washington; Geneva, New York; and Experiment, Georgia. In 1956, Congress authorized $450,000 to build the National Seed Storage Laboratory in Fort Collins, Colorado.

According to Dr. Steve A. Eberhardt, the national lab's director since 1987:

> *Until then when people did realize that seeds contained characterisics worth preserving, the seeds were sent to the federal Bureau of Plant Industry outside Washington, or to research workers in state experimental stations. But neither of these groups had adequate storage facilities. If the seeds had no obvious outstanding attributes, they were either discarded or tossed in a drawer and left there until all viablility was lost.*

The corn blight disaster of 1970 shocked policy makers into creating a more formal network. The National Plant Germplasm System was soon formed with facilities including ten clonal germ plasm repositories for plants that can't be reproduced from seed, several crop-specific seeds and genetic stock collections, and a laboratory and quarantine center at Beltsville, Maryland.

The National Seed Storage Laboratory is said to be the biggest seed bank in the world, but it is still lacking. After decades of underfunding, however, it recently received a shot in the arm, receiving funding for a long-awaited expansion.

Completed in 1958 and operated on a budget of about $2 million a year, the National Seed Storage Laboratory (NSSL) consists of a nondescript three-story yellow brick building housing offices, laboratories, and nine storage vaults that are accessible from a common corridor. The vaults hold more than 250,000 seed samples representing over 600 genera and 1,300 species, including all major crops grown in the United States.

Most of the seeds in the collection come from public service agencies, primarily from the federal plant introduction office. The lab is most anxious to obtain wild species and landraces, such as corn from Central and South America, before their centers of origin are destroyed by development. Gene banks in other countries sometimes use the American lab as a backup for their collections. Other depositors to this national seed bank, strictly on a voluntary basis, include commercial seed firms and individuals involved in plant breeding or in seed research.

According to Dr. Eberhardt:

> *Contributions from seed companies and breeders include newly released cultivars, those approacing obsolescence, and those already obsolete. The corn blight proved the value of the latter group. No one can tell if a cultivar that is susceptible to certain existing pathogens would not be resistant to a new race of pests that might show up in the future. We frequently receive requests for seeds of cultivars that have been missing from the agricultural community for thirty to fifty years.*

The needs of farmers can change because a new pest accidentally enters the country, because a disease-resistant strain loses its immunity, or because of the greenhouse effect. But more frequent and less dramatic are the demands brought about by cyclical and expected changes in weather patterns. In North Carolina, for example, farmers have been facing periods of drought for the past few years. They are trying a new tropical corn developed for Central and South America that is more tolerant of insects, diseases, and heat compared with the varieties presently grown. This is the second year of raising about three thousand acres of tropical corn, planted after the wheat and barley harvests in mid-June.

Dr. Loren Wiesner, NSSL curator, said the lab also accepts heirloom seeds from individuals who have a variety not already in the collection. The only requirement is adequate documentation, consisting of descriptions of the varieties or the breeding lines or references to publications in which such descriptions can be found. Because few gardeners have such documentation on seeds passed down within a family or a community, there are few American heirlooms in the NSSL collection.

Nor can home gardeners make withdrawals from the bank. The lab exists to preserve seeds, not to provide them. Individuals wanting seeds to grow will probably be directed toward one of the private seed savers groups. "We're not a procurement agency," said Dr. Eberhardt, "but we will advise individuals as to sources of seeds stored in the laboratory." But any bona fide research workers in the United States or its territories can obtain nominal amounts of seed without charge if supplies are not available elsewhere.

Before storing the seeds, scientists test them for viability by placing them on moist paper towels. This paper toweling, specially made for germination testing, is rolled and then placed on end in stainless-steel chambers that keep the seed at an optimum tempertature. If a majority of the seeds germinate, the remaining seeds are placed in drying rooms, where their moisture content is reduced to 5 percent or 6 percent—about half the usual amount.

After drying, the seeds are hermetically sealed in moisture-proof aluminum foil bags that weigh about seven ounces. These are arranged in numbered steel trays, filed in numbered steel racks, and stored in typical cold storage rooms with temperatures kept at –18°F. Conventional refrigeration methods are generally used or special cryotanks cooled with liquid nitrogen at –196°F. Such frigid temperatures actually slow the metabolism within the seed to such a point that they hardly age at all.

For most accessions, such as seeds of named varieties or of self-pollinating plants, the lab usually stores about fifteen hundred seeds. With varieties of cross-pollinating plants, they store three thousand seeds. Because only a low percentage of hybrid seeds reproduce the qualities of their parents, the NSSL does not store them but attempts to obtain each of the parents used to produce such hybrids.

According to Dr. Wiesner:

> *The regional stations like to have about 7,000 seeds, because it's from those seeds that distributions are made to researchers and breeders. However, they lack the refrigeration facilities to store seeds for long periods. The regional stations' working collections have been compared to checking accounts in relation to the NSSL's base collections, which are long-term savings accounts. The base collection serves primarily as a reserve in case the working collection is damaged or destroyed.*

But there are also some differences in the types of seeds found in each. In Colorado, the largest quantities of seeds in storage are in base collections of small grains, cotton, soybeans, maize, sorghum, and tobacco. Each regional station specializes in certain types of plants, depending on climate and the economic importance of a crop to its region. If you wanted to donate tomato seeds to the system, you would be referred to Geneva. Beans would go to Pullman and cantaloupes to Ames, whereas okra would be sent to Experiment, Georgia.

With all the undertakings of the U.S. government, there are critics devoted to The National Seed Storage Laboratory. Most complaints usually concern the storage facilities. For example, not all the seeds at the regional station are backed with base collections in Colorado. Dr. Wiesner calls this goal a "first priority."

In 1988, Associated Press science writers Lee Mitgang and Paul Raeburn wrote an award-winning series in which they reported that many of the seeds held by NSSL were dead or dying. Jeremy Rifkin, known best for his stand against genetic manipulation, filed a suit charging that gross negligence on the part of the National Plant Germplasm

System was harming the environment. The latest criticism, from the National Research Council, revolves around the system's "cumbersome administrative structure," and similar to nearly all these critiques the solution boils down to a lack of funding. Seed storage and collection is so low in political priority that it doesn't even have its own line item in the federal budget, and there are no Capitol Hill staff members assigned to the issue.

Far from being blamed for NSSL's problems, Dr. Eberhardt and Henry L. Shands, head of the overall germ plasm program, are credited with several recent improvements. In 1988, the lab's budget doubled to its current $2 million level. The laboratory's computer system, the Germplasm Resources Information Network, is said by researchers to be keeping much better track of the system's inventory.

But the most pressing need has been space. The storage vaults in Fort Collins are so full, according to Dr. Wiesner, that eight major crops are being stored at regional labs rather than in the vaults of the NSSL. Expansion plans include increasing storage space from five thousand to twenty thousand square feet.

Yet bricks and mortar aren't the only things needed for the seed storage system. The institution needs both staff and physical space at the regional stations to ensure each line of germ plasm in the NSSL vaults is kept alive and unaltered.

"Regeneration of seed lots with low germination is a continuing need," read a report from the National Plant Germination System called *Managing Genetic Global Resources.* "A large proportion (almost 50 percent) of the accessions at NSSL are below the minimum desired size (550 seeds). Regeneration of these samples is urgently needed."

Every five to ten years the seed is retested. If viability drops below 60 percent, the appropriate regional curator is responsible for producing a new generation of seeds with the same genetic composition as the original accession. This is easier said than done.

Like people, seeds carry many genetic traits—some good and some bad—but not every seed holds every desirable attribute. And as seeds age, their ability to germinate begins to fail, eventually reaching a point at which they will no longer respond.

If a seed sample has declined with age, genetic diversity can be compromised. Which of the seedlngs that result from these aged seeds best represents the parents? This decision is not too difficult when dealing with the color or size of fruit, but what about the changes that are not readily apparent? What if the plant that results from weakened seeds can no longer withstand a certain number of hours of below-freezing temperatures or live three weeks without water when previous seedlings faced up to these travails with ease? So the very act of choice leads to a potential genetic drift and the danger of losing many small but, nonetheless, important individual traits in the plant's genetic history.

But if deterioration rates are only allowed to fall to 10 percent before new germination is required, as many critics have suggested, the samples will be disturbed more frequently and more seeds lost simply by the act of handling. The growing out process itself can alter the genetic material. Ideally, seeds should be grown under conditions similar to those where they were native, but because seeds are collected from around the world, that isn't always possible. Once again, there is a problem of genetic drift.

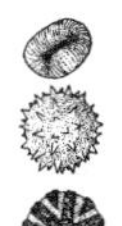

Call for update from ARS-USDA, National Seed Storage Laboratory, Colorado State University, Fort Collins, CO 80523. Telephone: (303) 484-0402. Business hours are from 8:00 A.M. to 4:30 P.M., Monday through Friday except holidays.

THE VIEW FROM ENGLAND

The idea of collecting seed as a hedge against future calamities and weakened germ plasms is not limited to the United States. In addition to the more than five million preserved plant specimens in the collections of The Royal Botanical Gardens at Kew, Wakehurst Place, the sister institution of Kew, is home to England's collection of seeds.

An hour and a half away by car from Kew and just a few miles from the seaside city of Brighton is the five-hundred-acre rural estate of Wakehurst Place. High on the Sussex Weald, Wakehurst Place boasts an average rainfall of 32.3 inches a year, a wide-ranging topography, and a romantic Elizabethan mansion. There, in a deep freeze, is another collection of seeds. Safely filed away in color-coded, numbered drawers are layer upon layer of aluminum phials and glass jars, containing seeds of nearly four thousand species, or 1.5 percent of the world's flowering plants.

The seeds stored at these facilities are called orthodox seeds. Like their American cousins, these seeds can survive desiccation to the point that water comprises no more than about 5 percent of their weight. If they are kept at –20°C, physiologists predict their storage lives to be measured in centuries. Germination is monitored every ten years.

Most seeds in the British Isles are orthodox, but there are exceptions called recalcitrant seeds: Chestnuts, acorns, and samaras (the one-sided, winged fruit of the sycamore) die when they dry out, so they cannot be stored for any length of time. Also, many tropical trees have seeds that are large and fleshy and exhibit recalcitrant traits.

The physical setup of the Kew Seed Bank consists of a suite of three –20°C cold rooms (each approximately two thousand square feet), a dry room with an air lock held at 15°C and 15 percent relative humidity, as well as cleaning, X-ray, and general laboratories. Recently, an additional drying room was added. Germination equipment is shared with the research group and includes a wide range of alternating and constant temperature incubators with a range of 2°C–36°C.

When seed arrives, it is immediately placed in one of the two dry rooms. At 15°C and 15 percent relative humidity, seeds can dry out under the best possible conditions for retention of long-term viability, making them suitable for cleaning.

The cleaning of seed requires great care, and it is important to use a microscope to check fruit samples for damage after they have been removed from the seed head. Aspirators are used to blow off chaff and empty seeds. Because it's a messy business, the operation is carried out within clean air cabinets linked to a filtration system.

A sample of clean seed is then examined using X-ray analysis. This is to check both the contents of the seed and, in more complicated dispersal systems, to check number of seeds that are present.

Once the seeds are cleaned, the total numbers are counted. Five samples of fifty seeds are weighed to provide a mean with a standard margin of error. The count is deliberately underestimated on the computer, so it is unlikely that a collection will be depleted by overdistribution.

Cleaned seeds are then given a period of at least one month in the 15°C, 15 percent relative humidity room. During that time, they equilibrate down to between a 3.5 percent and 6.5 percent moisture content. It is also important when drying cleaned seeds that they are well aerated and neither kept in a dead spot within the dry room nor too tightly packed within cloth or paper bags. If possible, the seeds are spread in a thin layer. When drying is complete, a reading close to 15 percent relative humidity is needed.

Depending on their bulk, dried collections are placed in a variety of containers. Because the seeds must be accessible, screw-top containers with a good compound seal are best. Very large collections are stored in fruit-preserving jars.

It's amazing to realize that one room at the Kew Seed Bank holds almost 8,600 accessions of about 3,700 species. Dr. Hew D. V. Prendergast, the Overseas Seed Collector for the Kew Seed Bank, had this to say of the collection:

> *The 4,000 or so species in the Seed Bank reflect the two main collecting programs of Kew's Seed Conservation Section: the floras of the United Kingdom and those found in the arid and semi-arid tropics. We save plants under threat. One of our recent targets has been the population of clove-scented broomrape* (Orobanche caryophyllacaea) *growing at*

the site of the 1993 British Open Golf Championships, Royal St. Georges!

And the interest is not just with the rare. So far, some 550 species of the native British flora, including, for example, the nationally extinct grass *Bromus interruptus*, have been banked. Along with seed collections from more than one hundred other countries, these now compromise the world's most diverse store of wild-plant germ plasm.

Kew is the first to admit that nobody knows just how long seeds of most wild species will survive. In some respects, the Seed Bank can be viewed as one huge experiment in aging with each ten-year retest offering valuable data. Studies to date suggest that a two-hundred-year storage life may be widely achievable in the grass family. But just to monitor a collection's viability over this time frame would use up 750 seeds. Assuming that the initial test can break dormancy within two tests of fifty seeds, ten yearly monitors are carried out with fifty seed tests over the first sixty years, and twenty yearly monitors thereafter—thus, the Seed Bank policy of collecting at least twenty thousand seeds from the wild for any species of plant.

"For the last five and a half years my task as Overseas Seed Collector has been to plan and execute expeditions in search of useful plants, rather than unusual ones," said Dr. Prendergast. "Ironically, it is often the most useful plants which are most vulnerable to overexploitation by people and livestock; the most termite-resistant trees are the first to be chopped down for house building and the tastiest herbs are the first to be grazed out. On top of tangible threats like these is the danger posed by climatic change."

Kew has decided to emphasize the dry tropics because they do not have the romance associated with the blousy orchids or vining treasures of the rain forest. And when they do burn, it is with far less drama than the smoking ruins of Amazonia. But these areas cover far more land and support millions more people than the forests do. And unlike many of the rain forest plants, the seeds of dry tropical plants are orthodox and easily adapt to freezing and long-term storage.

Dr. Prendergast said,

My first trip was to Australia where rainstorms had fallen with as much rain in a few days as Central Australia usually gets in a year. The famous red sands were ablaze with color. Among the annuals, mulla-mullas, Ptilotum (Amaranthaceae) *made pink carpets and numerous daisies such as* Helichrysum *endlessly lined the desert roads and tracts with whites and yellows. What interested me more, however, were the fruits that followed the flowers: The shiny black ones of* Solanum *(those of several species are edible), the pods of leguminous*

> Senna *and* Indigofera, *and the winged or spiny fruits of local equivalents of our native goosefoots, such as* Atriplex. *Australian members of this genus are important fodder plants widely introduced elsewhere, especially to North Africa. As in subsequent expeditions, the grasses were most prominent among my collections. Some, already of great local interest, could also have a bright future in other dry regions, but are difficult to obtain. The role of the Kew Seed Bank is simple: Store the seeds well, publicize their presence in an annual seed list, and make them freely available for investigation.*

After four vists to Oman, Dr. Prendergast has narrowed his sights to the southern province of Dhofar, floristically the richest and the most threatened region of the country. Dhofar's mountains are covered from June to September by monsoonal mists and rain, but many of the local plants are in a decline as a result of recently acquired oil wealth.

Said Dr. Prendergast,

> *I have concentrated on collecting seeds from the trees on which fog moisture condenses. The water dripping off supports a dense ground layer of herbs and grasses and recharges the groundwater reservoir on which the coastal towns and gardens depend. In the Seed Bank there are now some 13,000 seeds of* Blepharispermum hirtum, *an endemic composite shrub, 7,400 seeds of the endemic euphorb* Jatropha dhofarica, *and still uncounted collections of the dominant Dhofari tree* Anogeissus dhofarica *and* Lawsonia inermis, *whose leaves are the source of henna and whose seeds I have collected in the arid southwest of Madagascar.*

In May of 1993, when lowland temperatures were exceeding 113°F, Dr. Prendergast spent two days at some seventy-two hundred feet on the main massif, Jebel Akhdar, or the Green Mountain. There he collected the usually five-seeded fruits of *Juniperus excelsa* and remembered waking before dawn to the cooing of wood pigeons. Even more so than the juniper, this bird has been restricted to the highest places of Oman by a gradual warming and drying of the climate. A few thousand feet lower, a wadi bed was lined by a small, spiny tree called *Sideroxylon mascatense,* the only Omani member of a family usually restricted to the wet tropics. Described in the ninth century and still known by its classical Arabic name of *but,* this tree has the most delicious cherrylike, but raisin-flavored fruits whose taste made them easy to collect. It would, thought the doctor, make a good candidate for a trial as a food source in arid climates elsewhere.

Offering up seed for trial comes around once a year when the Seed Bank's *List of Seeds* is published. The edition for 1992 had seventy-nine pages and offered seed based on the following requirements: The seeds must be used for the common good in the areas of research, trialing, breeding, education, and the development of public botanic gardens.

If the recipient seeks to commercialize either the germ plasm or its products, or research derived from it, then permission must be sought from the Royal Botanic Gardens at Kew. Such commercialism will be subject to a separate agreement complying with Kew's policy that a proportion of net profits be distributed to the country from which the seed was collected. Finally, the germ plasm must not be passed on to a third party for commercialism without the permission of the garden.

Then come the pages listing twelve species of *Campanula* collected in Turkey, Austria, Bulgaria, or Scotland; *Petrorhagia prolifera*, or the Childing pink, an annual collected throughout Greece, Yugoslavia, Italy, or Bulgaria; or hundreds of species of grasses, collected in New Zealand, Botswana, Indonesia, India, or Greater London.

In 1936, Hermann Goering said in a German radio broadcast: "Guns will make us powerful; butter will only make us fat." Fifty-five years later, the arguments between funding guns or funding butter still persist. With the building of the new seed storage facilities in Colorado and with the collecting taking place at Kew, the scales have tipped slightly toward feeding the world rather than feeding the pile of munitions. But for how long?

Note: Visitors are welcome at the NSSL, but the staff asks that tours be arranged as far in advance as possible. If interested, call or write the Research Leader, ARS-USDA, National Seed Storage Laboratory, Colorado State University, Fort Collins, CO 80523. Telephone: (303) 484-0402. Business hours are 8:00 A.M. to 4:30 P.M., Monday through Friday except holidays.

Malabar Gourd

CHAPTER 11

The Big Business of Seeds

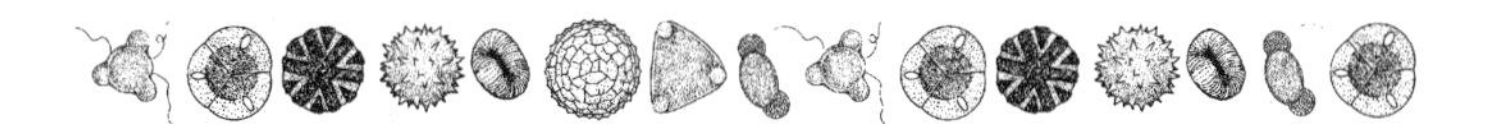

Late in 1991, Procter & Gamble decided to switch from using 100 percent soybean oil in the manufacture of Crisco oil to a blend of canola, sunflower, and soybean oil. This broke a long-standing tradition of using only soybeans and threatened to create a shift resulting in the loss of a market for 225 million pounds of soybeans. To grow this amount of beans would require 620,000 acres of farmland.

DeKalb Genetics Corp., of DeKalb, Illinois, recently announced that improved results from its North American and international seed operations led to sharply increased earnings for 1994. End-of-the-year net earnings were $10.6 million, while sales rose to $28.9 million. Total revenues were up to $247.5 million in 1993 with sales volumes increasing for seed corn, soybeans, and sorghum.

Pioneer H-Bred International, Inc., Des Moines, Iowa, announced that its earnings for fiscal 1994 reached a record $213 million on sales of $1.479 billion. They cited sharp growth in seed corn as a significant factor in better earnings.

Potatoes were introduced in early colonial times and were grown in New England as early as 1719. Over the next one hundred years, breeding programs were limited, but new varieties began to show up on the market about 1840. Most of the best new varieties were developed from true potato seed. Great progress has been made in combating diseases like early blight, fusarium wilt, scab, and various viruses, but scientists at Calgene Pacific, Melbourne, Australia, recently may have found a potato capable of boosting yields by as much as 200 percent. Estimates for the current seed potato world market is $20 billion.

The Advance Seed, Krugersdorp, South Africa, recently came under new ownership by Whitnev Investments (Pty) Ltd., a British Virgin

Islands–owned holding company, originally established in 1948. Advance Seed is a contract grower, wholesaler, importer, and exporter of seeds, including grasses, millet, and sorghum for forage. Currently, the company has annual revenues of $18 million and hopes to increase revenues to $25–$30 million over the next ten years. Recent political changes in South Africa, including the election of Nelson Mandela, are opening new trade opportunities in agriculture.

With the trade embargo against Vietnam lifted, agricultural exports to that country are expected to rise—and if the U.S. share of this business reaches the 16 percent attained in other Asian countries—it could reach $250–$300 million per year.

Where will Limagrain be in the next ten years? Limagrain, a giant of European agribusinesses, is considered to be the world's third largest seed company with annual revenues of $550 million. Over the next five years, the main target of the company will be to increase earnings, profitability, and efficiency.

It should be noted that the International Federation of Seedsmen (the French acronym is FIS) met 29 May through 3 June, in Oostende, Belgium, with approximately 1,020 delegates from around the world in attendance. When all the official dignitaries were factored in, the total was 1,350.

And in the May 1994 issue of *Seed World* (the bible of the seed industry), an international guide to exporters and importers was published. The list began by noting the companies included in the tally were involved with seed, seed equipment, or seed services. Over 150 companies were included, ranging from Illinois Seed Foundation, Inc., in Champaign, Illinois, to Samen Mauser AG, of Switzerland, Sheetal Hybrid Seeds Co., located in Jalna, Maharashtra, India, to Zeraim Seed Growers Co., Ltd., of Gedera, Israel.

The reports from these companies are just the tip of the iceberg. As a growing world population needs more food and more markets are opened on the international business front, the future of seeds is tremendous. Just like American baseball supports dozens of allied industries from bubble-gum cards to souvenier programs, seeds support a number of satellite industries.

Seeds are truly big business!

A FEW PRODUCTS

The Seedburo Equipment Company of Chicago is marketing a microprocessor germinator that features a forty-test program memory. The all stainless-steel germinator features a self-contained humidifier and holds up to forty trays with one and one-half inch spacing and provides automatic hot, cold, light, and dark testing capabilities.

Agriculex Research of Ontario just introduced an improved version of the corn sheller. Three rollers covered with rubber-type, wear-resistant materials reduce kernel damage, and although it is designed mainly for single-ear shelling, it can also shell bulk samples.

The Redwood Empire Awning Company markets pollination cages of galvanized steel frames with mesh covers that feature vinyl-coated nylon mud flaps and corners. They guarantee that no pollinating insects can get through.

There are also companies that specialize in making bags to store or ship seeds, custom printers for labels, computerized programs for making individual labels, manufacturers of paper, plastic sheeting, burlaps, along with all types of sewing equipment to seal the bags when they are full of seeds. And for the garden market, envelope maufacturers produce seed packets made of white Kraft paper, either plain or printed, in one to four colors, some with special viewing windows, and self-sealing gum flap seals.

Huge machines are now available to apply protective coatings to individual seeds, often treating three thousand pounds of seed in one hour. There are coatings of pesticides and fungicides; coatings to protect against microorganisms; colors sprayed on for identification; sealers to control the amount of moisture that touches a seed; and pelletizing, the process of coating tiny seeds so they are easier to handle.

In addition to products, as of late in 1994, there are more than thirty-five independent seed testing laboratories in the United States and Canada. Their services include testing for germination, vigor, cold, accelerated aging, cultivar purity, genotyping, and varietal identification.

NEW PLANTS

In the March 1995 issue of *Seed World* the following new varieties were announced:

Zajac Performance Seeds' newest turfgrass varieties are 'Nordic' hard fescue and '18th Green,' a creeping bent grass that was developed to withstand winter temperatures and snow.

Four new peppers join the new vegetable lineup at Rogers Seed Company of Boise, Idaho. The colorful new peppers include 'Mandarin', a glossy orange pepper with five- to six-inch fruits maturing from green to a deep pumpkin orange in seventy-four to seventy-eight days. Their new jalapeno pepper, 'Firenza', produces very pungent, dark green peppers in sixty-two to sixty-six days.

Harris Moran of Pleasanton, California, reports a new zucchini called 'Tigress', ideal for the eastern United States. The variety is a high-yielding, early-maturing zucchini with medium green color fruit.

N. L. Chrestensen of Erfurt, Germany, has released several new flower, vegetable, and herb varieties for the 1995 market. A new *Begonia semperflorens* called 'Ergo Dark Leaved', has bright single flowers of red, soft rose pink, and white; all are especially attractive against dark leaves. They also introduced two new cutting asters called 'Bornella', with dark rose flowers and yellow centers, and 'Migella', bearing red flowers with yellow centers.

Burpee announced 'Silver Choice', a seventy-five-day white sweet corn featuring eight-inch ears with sixteen rows of sweet, white kernels. The F_1 hybrid (see page 34), features strong, dark green husks. They also announced two new winter squash varieties: Lakota (*Cucurbita maxima*), a near-Hubbard type noted to mature in 105 days. The five- to seven-pound fruits have smooth, thin skins. 'Hasta La Pasta' (*Cucurbita pepo*) is an eighty-day, F_1 hybrid noted to produce seven- to eight-inch deep orange fruits high in beta-carotene. Like Lakota, this hybrid is said to have excellent eating qualities.

Farmers Marketing Corp. of Phoenix, Arizona, announced that they developed from seed a dense, fine-textured Bermuda grass called "Princess."

The lawn seed business alone continues the development of new grass varieties every year for a country that is seemingly wedded to conservation. Ironically, we continue to pour millions of dollars and millions of man-hours into the development of an artificial convention that was never meant for a country like the United States.

The Lawn Seed Division of the American Seed Trade Association (ASTA) met during ASTA's annual convention in June of 1994 in Minneapolis, where they reported on the consumption of various lawn seeds. Included in the news were reports that red fescue sales were up 20 percent to 25 percent in 1993, and Kentucky bluegrass consumption was up 5 percent over the previous year. The increases were attributed to overseeding by landscapers, seeding areas around new housing construction, and seed to replace winter damage from the previous year. At the same time, sod growers reported their sales up 15 percent over 1993.

John Glattly of Twin City Seed Company noted that new home construction in Minnesota was having a very positive effect on the consumption of turf grasses, especially for Kentucky bluegrass. Three hundred new golf courses in Minnesota alone were also good for the grass seed business.

From the flower's point of view, the All-America Selections (AAS) award-winning black-eyed Susan (*Rudbeckia hirta*) cultivar for the 1995 market is 'Indian Summer'. Here's a case of taking a time-honored American wildflower that grows throughout the country and has not exactly meant high fashion in most American gardens and bringing it up to date.

Flecke Saaten Handel of Wunstorf, Germany, took this common, some would say weedlike flower, and turned it into a plant that bears long-stemmed, single or double blossoms of golden yellow that measure six to nine inches in diameter. The plants reach a height of three to three and one-half feet in the garden and will flower about ten to twelve weeks when grown under late spring conditions. The new AAS flower will be sold through mail-order catalogs and in seed packets in many stores. The qualities of this new flower are considered to be so good that judges from AAS awarded the plant "fast track" status, enabling the owners to get marketing and advertising programs to seed company buyers early.

THE FOOD MARKET

Today, white corn is one of the fastest growing commodities in the American snack-food industry. It's used in tortillas, chips, and other fast foods. But white corn products are no better (or worse) than yellow corn. Yet the same reasoning that makes consumers think white flour is superior to unrefined brands and that white eggs are superior to brown eggs leads to the belief that white corn products are better than yellow.

In 1992, the United States produced about 570,000 acres of white corn compared with about 73 million acres of yellow corn. But when used for food, white corn gets between $.50 and $.75 more per bushel than does yellow corn.

At one time, white corn was considered a low-yield crop that was difficult to manage, but improved breeding has led to the white kernels competing favorably with their yellow relatives. Today, it's fairly close to yellow corn when it comes to yield. Genetically, the two forms are very similar, differentiated by a white endosperm in the white corn.

Most farmers contract directly with the food processor. And the processors are very exact about the qualities they demand. The food industry wants superior grains and early harvest. To ensure high quality, the food industry has set standards originally established by the Texas Corn Producers Board.

Today, white corn is the fastest growing snack-food ingredient. This year's biggest crop producer was Texas, which produced nearly 13 million bushels on 114,000 acres, followed by Nebraska and Kentucky.

Before the fall of the peso, the North American Free Trade Agreement was going to have a positive effect on U.S. exports because if

Mexico became more affluent, the demand for white corn was likely to increase.

Projections for the 1992–93 crop year showed that Venezuela would be the top buyer of white corn from America, importing nearly 188,000 metric tons, South Africa would be second, with 120,304 metric tons; and Japan third, with imports of 85,500 metric tons.

OVERSEAS MARKETS

The market for seeds is unlimited. As populations grow and hunger spreads, many countries are interested in expanding their seed businesses.

China Wants to Learn

A delegation from China attended the ASTA Vegetable and Flower Seed Conference in January 1993. Among those in attendance were representatives from Liaoning East-Asia Seed Corporation, which is based in Shenyang City, People's Republic of China. They were interested in establishing cooperation with American firms to learn as much as they could about vegetable seed breeding and updating production methods. Cheng Guang Tao, the general manager of the company, told the Americans that his company can economically produce tomato, cucumber, watermelon, pepper, onion, and lettuce seed because of low labor costs. He suggested that Liaoning could multiply parent seed from the United States and then send it back.

Liaoning is also looking for American and European varieties that might be suitable for the Chinese market. Tao told Americans that his company has about one hundred sales representatives located throughout China, with about sixty working in Liaoning Province, located in the southern part of Manchuria.

There are approximately 4 million hectares (a hectare equals just a bit less than two and a half acres) of land under cultivation in Liaoning Province. The province is located in the northern Temperate Zone with average temperatures that range from 5°C to 10°C. It's the largest producer of vegetable seeds in China. Their prized cultivars include three varieties of tomatoes, two eggplants, three peppers, and a medium maturing watermelon that matures in ninety-five to one hundred days.

The View from Holland

Barenbrug, the Oosterhout, Holland–based fodder crop and grass seed breeding and marketing company, is once again an independent firm. For three years it was a holding of the British multinational Unilever. About a year ago, Unilever began to sell its agricultural businesses, and

Barenbrug bought back his company. He thinks that more multinationals will sell their seed divisions in the 1990s. "A lot of multinationals have gone into the seed business based on the nice stories which have been told about all the possibilities with biotechnology. Today, everyone looks more realistically at those possibilities," said Barenbrug.

"The seed business is very long term," he said. "Some multinationals are not used to that. Very often, they cannot obtain much synergy. Profits in the seed trade are very variable. In general, multinationals do not like that."

Although Barenbrug is once again independent, it will collaborate with Unilever's plant research unit, Plant Breeding International (PBI). Barenburg noted that, "The PBI crops we will represent will be different per country. But these will be mainly cereal, oilseed rape, and pea varieties for which we have exclusive sales rights in a number of countries.

"Our national subsidiaries were responsible for only their national business. Now we will see which ones can do the best business in different countries."

The result is that Barenbrug USA will become responsible for seed sales to the Pacific Rim, Mexico, and Canada. And the company plans to remain flexible since changes happen so fast in the industry.

AMERICA'S BIGGEST DEALERS

And who are the largest seed companies in America? The following are gleaned from companies that agreed to provide information to *Seed World*. It includes many companies with overseas branches. Ciba-Geigy, for example, is Swiss, and there just might be others.

Agrigenetics Company is located in Eastlake, Ohio. They sell seeds of corn, soybean, sorghum, cotton, alfalfa, and sunflower. In 1990, their annual seed revenues were $117.2 million. The number of full-time employees is 750.

Asgrow, a subsidiary of The Upjohn Company, resides in Kalamazoo, Michigan. They sell farm seeds, corn, vegetable, sorghum, soybean, alfalfa, and sunflower. Asgrow worked with Monsanto Company and E. I. DuPont de Nemours and Co., Inc., on a herbicide-resistant corn and soybeans. Approximate sales in 1990 were $276.75 million.

Geo. J. Ball, Inc., sells vegetable and flower seeds. It's a family-owned organization with subsidiaries in many areas of the commercial horticultural world. Some of these offshoots are the Pan-American Seed Company, W. Atlee Burpee Co., and Geo. J. Ball Publications and *GrowerTalks.* Their annual seed sales and the number of full-time employees are confidential.

Biotechnica International, Inc., is located in Overland Park, Kansas. They sell hybrid seed corn and soybeans. The reported revenues for the period ending June 1991 was $12.3 million.

Calgene, Inc., is in Davis, California. They sell cotton, canola, potato, tomato, and soybean seeds. In 1991, their annual sales were $21.5 million.

Cargill Hybrid Seeds of Minneapolis, Minnesota, deals in hybrid corn, sorghum, and sunflowers seeds, and on the international market, hybrid wheat, rice, safflower, and soybeans. Annual sales are confidential, but they have sixty thousand employees worldwide, with thirty-four thousand employees in America.

Ciba-Geigy Seed Division is located in Greensboro, North Carolina. They sell corn, soybeans, and alfalfa.

DeKalb Genetics Corporation is in DeKalb, Illinois. They sell hybrid corn, sorghum, sunflowers, forage sorghum, varietal soybeans, and alfalfa.

DowElanco of Indianapolis, Indiana, deals in herbicides, insecticides, fungicides, and seeds. When Dow and Eli Lilly Company joined in 1989, total projected sales were $1.5 billion.

The Genesis Group, Salt Lake City, Utah, sells forage and turf grasses, farm seed, wild bird food, vegetables, and wildflowers.

Jacklin Seed Company, Post Falls, Idaho, is a researcher, producer, and marketer of turf, foliage, and low-maintenance grass seeds. Revenues are confidential, but they support more than one hundred full-time employees.

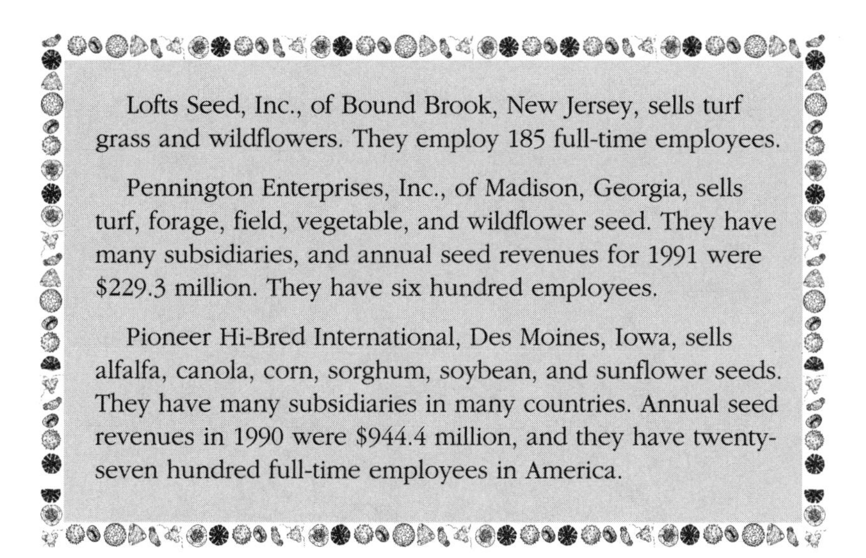

Lofts Seed, Inc., of Bound Brook, New Jersey, sells turf grass and wildflowers. They employ 185 full-time employees.

Pennington Enterprises, Inc., of Madison, Georgia, sells turf, forage, field, vegetable, and wildflower seed. They have many subsidiaries, and annual seed revenues for 1991 were $229.3 million. They have six hundred employees.

Pioneer Hi-Bred International, Des Moines, Iowa, sells alfalfa, canola, corn, sorghum, soybean, and sunflower seeds. They have many subsidiaries in many countries. Annual seed revenues in 1990 were $944.4 million, and they have twenty-seven hundred full-time employees in America.

The multiple flower of the salmon blood lily (Haemanthus multiflorus) *bears up to one hundred individual blossoms, all eventually ripening to seed-filled pods.*

CHAPTER 12

A Sampling of American Seed Houses

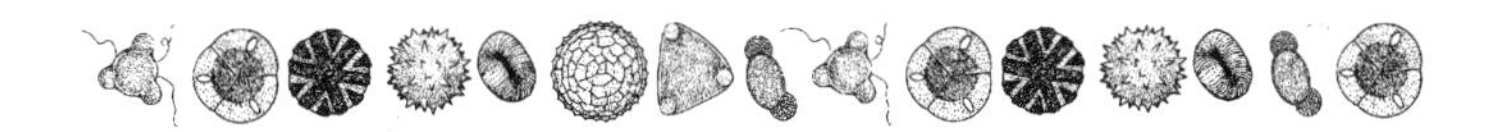

JOE SEAL OF BURPEE SEED

Joe Seal owned The Country Garden, one of the most imaginative seed houses operating during the 1980s. For a few years during the early 1990s, he was at large in the seed world but finally settled at Burpee Seed Company, one of America's first and most famous seed companies, founded in Philadelphia in 1888.

> *W. Atlee Burpee was a drop-out from the University of Pennsylvania Medical School at the age of eighteen, trading a career in medicine for a career in the world of horticulture. In 1876, he borrowed $1,000 from his mother, and founded a seed company that would find new seeds, develop new sources, and in the end, sell only the best seeds that would grow in America's diverse climate and never compromising on quality. He guaranteed satisfaction for one year from the date of the order, and if you weren't pleased with your seeds, either he would replace your purchase or send you your money back. I might add, it was amended in 1985 and now there's no time limit to the guarantee.*
>
> *The company was successful and in 1888, Mr. Burpee established Fordhook Farms in Doylestown, Pennsylvania, as an experimental farm that would test and evaluate new varieties of both vegetables and flowers, and, of course, produce seeds. This experimental testing station at Fordhook was started before the United States government had a seed testing or even a research station, although the U.S. Department of Agriculture was formally established in 1862. In the early 1890s, the company's slogan became "Burpee seeds grow."*

But Mr. Burpee wasn't content to remain in Pennsylvania. By the 1890s he began trips to California, looking for a warmer spot to produce seeds, and in 1909, he established Floradale Farms in Lompoc, California. Lompoc, you may recall, was the town that W. C. Fields made famous in his movies, and today, it's still famous for being a fertile valley, reknowned for its flower production and steady seeds commerce. This was especially important because at the turn of the century, the sweet pea was the favorite annual flower grown in American gardens across the United States.

According to Alice Morse Earle, by the early 1900s, sweet peas were favorites in American gardens for well over a century. They were grown in cottage borders and formal gardens, sold in the cut flower market, and carried in nosegays and posies. It's interesting to note the change in flower fashion. Sweet peas are still one of the most popular flowers in English gardens, and today, England's Thompson & Morgan devotes two full pages to these flowers, listing over thirty cultivars. In comparison, Burpee's catalog offers only three cultivars: Burpee's 'Galaxy Mixed', 'Patio Mixed', and 'Pink Perfume'.

By the time W. Atlee Burpee died in 1915, his seed company was the largest mail-order seed company in the United States, and possibly the world. During his lifetime, the following popular vegetable cultivars were introduced: Burpee's 'Surehead Cabbage' in 1877, Burpee's 'Iceberg Lettuce' in 1894, and their stringless green pod bush bean, as well as the world famous Fordhook bush lima bean in 1907.

> *In 1915 at the age of twenty-two, David Burpee, W. Atlee's son, became president of the company and remained at the helm for the next fifty-five years. During his tenure, he charted the company's course over two world wars, the Great Depression, in fact all the way from the age of the buggy whip to the moon landings.*
>
> *Some of the varieties that Burpee introduced during that time were the red and gold hybrid marigold, the first hybrid flower from seed to be offered for commerical sale in the United States. This flower was followed by the 'Giant Climax' hybrid from 1957, the everblooming 'Nugget and Fireworks' marigolds, in addition to the 'Lady Hybrid' marigolds. Burpee's first 'Lady Hybrid' was an All-America Selections Winner in 1968 and voted an all-time favorite by home gardeners in 1984.*

Amazingly enough, in 1995, the 'Lady Hybrid' and the climax hybrid series are still being offered but have now been expanded into a number of colors. Also of particular merit is Burpee's 'French Vanilla', the first hybrid white marigold, and for gardens with afternoon shade, 'Snowdrift'.

David Burpee spent a lifetime both developing and promoting a white marigold. For twenty-one years, Burpee offered ten thousand dollars to the first person submitting seeds that would produce marigolds with white flowers, at least two and one-half inches across. On 28 August 1976, Mr. Burpee presented the check to Mrs. Alice Vonk for her white marigold.

David Burpee's most zealous crusade began in 1959 with the promotion to make the marigold America's national flower. In 1683, J. W. Gent wrote in *The Art of Gardening*: "There are divers sorts besides the common [marigold], as the African marigold, a Fair bigge Yellow Flower, but of a very Nauthty Smell." When you consider the marigold's reputation, Mr. Burpee's campaign had a hard row to hoe.

Said Joe,

> *Burpee loved flowers but he knew that vegetables needed a great deal of improvement if they were to be continually produced in American gardens. His hybrid cucumber and the Fordhook hybrid tomato, were the first garden vegetables, except for sweet corn, to be offered for commericial sale. These were the vegetables that helped American's victory gardeners of the Second World War, producing the bumper crops in backyards. These hybrids led the way to many of the newer vegetable varieties, including the most famous tomato of them all, Burpee's "Big Boy," introduced in 1949 and still being grown today. It has been joined by "Early Girl," "Better Boy," and "Big Girl."*
>
> *David Burpee also scored a big hit with the first patented vegetable grown from seed. This was "Green Ice," an open-pollinated variety introduced in 1973, after his retirement from the firm, and was awarded the first Plant Variety Certificate from the United States Department of Agriculture for its unique qualities.*

JAN BLÜM OF SEEDS BLÜM

Jan Blüm looks like the heroine in an Andrew Wyeth painting, preferably one that features the mountains of southern Idaho as the background to a dusty road that leads to a path strewn with poppies, winding up at an old farm building bearing her name. Seeds Blüm, pronounced "bloom," a name that, if it wasn't her own, would lead you to suspect that a press agent chose it.

With long brown braids and a cheery smile, Jan leads a business devoted not only to marketing, but also to collecting, preserving, and propagating millions of heirloom seeds, while shepherding a growing business when many seed companies are experiencing declines. And she still finds time to produce one of the most unique catalogs in the nursery industry.

> *I grew up in a central California farming community with parents who, though professionals, both grew up on farms. So as a child we had a big garden and my mother, not my father, took care of a magnificent lawn with nary a weed in it. I remember being enchanted with plants, but as a teenager, I wanted to do something with my life and not spend it weeding a lawn, so I married a minister and devoted myself to working with people.*
>
> *During the early 1970s, when everybody else was trying to get back to the land, we moved to Los Angeles. After a year in that environment, we heard of a landowner in Idaho who wanted a caretaker for 600 acres of pine trees with a creek running through them. Now who could refuse an offer like that?*

So they bought three acres and settled down to living a compromise: Jan wanted the wilderness, and her husband wanted a university town, and because they weren't far from Boise, it worked.

Then, with the help of a neighbor, Jan started a garden. He soon filled her head with the lore of vegetables, including how the old varieties were disappearing and blousy hybrids were taking their place. He gave Jan the descendants of some seeds he had been given as a wedding present in the 1920s, and she soon learned that flavor wasn't a thing of the past.

At the same time, Jan was teaching vegetarian cooking, and many of her pupils would ask the obvious: "What do we eat?" As a teaching aid, she filled glass jars with all types of beans and grains. It soon occurred to her that there must be hundreds of different grains and bean varieties. To fill the jars, she began to trade seeds with other gardeners, and before long there were over seven hundred different types of beans in the collection.

> *I really love food so I began to grow them out on the hill along with over 350 different kinds of tomatoes and I was hooked! Then I began to wonder why people weren't lined up to buy all these seeds. When super tomato sauce is available in your backyard, why are people still buying the dinky-dorf stuff found in cans? So I sent for all the seed catalogs I could find and began to order seeds only to find that the best of one year was gone to be replaced by a new hybrid by the next garden season.*

Jan read about the decline of plant germ plasm and incredibly convoluted political situations that contributed to the mounting problem. She began to think about doing something on her own. She wrote to other dedicated gardeners who were working to save the old varieties, including America's Kent Whealey of the Seed Savers Exchange and Lawrence Hill, the great English organic gardener.

Then, a local newspaper featured her garden, and the piece was picked up by the United Press International and sent all over the country. By the winter of 1981, she started her own seed company.

> *By this time I noticed, there were plenty of signals sent my way and when something hits you on several levels, that's when you can really commit to it. If it's only one level, it's usually not balanced enough and you end up leaving it behind!*

But the basic question usually asked is: "Where do the seeds all come from?"

> *We get them from all over the world. People send us seeds and most of our varieties come from our customers because they're trusting us to keep each of their favorite plants in existence. And we give many of our customers the chance to be our Seed Grower for a particular variety. After all, some of these gardeners have kept that favorite tomato or that beautiful annual flower or that disease-resistant squash for over thirty years so they obviously know what they're doing. And if they can't, we send the seeds to one of the hundred or so Seed Guardians. These are people who make a commitment to grow a certain variety and adopt that plant for a number of years.*

And that's just what a Seed Guardian is: Somebody who pledges to grow a chosen variety as long as he or she is alive. If something should happen to deplete the nursery stock, Jan can go back to a Seed Guardian for a new supply.

Sometimes though, they receive such a small amount of a chosen seed that they haven't enough to even try it out for taste or looks or productivity. So off it goes to a Seed Multiplier, a gardener who pledges to increase that particular plant until there's enough stock to go out to a number of gardeners around the country who then test it and let Jan know whether it's a hit or a miss.

> *After all, we're doing a lot of the work of the National Seed Storage Lab in Colorado, but instead of being located in one place, we're spread out over thousands of miles. And it's a lot of seed. If you have 20,000 volunteers, you can deal with 20,000 varieties! It's sort of like a Pandora's Box: You have a treasure chest and you really don't know what you have until you test it!*

The catalog for Seeds Blüm is just as unique as the business. "Doing the catalog allows me not only to work with something I love, but at the same time I can teach."

Almost one hundred pages feature vegetables, ranging from the Chinese yam (*Dioscorea batata*), a vine with a tuber that can reach two pounds in the first season (it was introduced in the United States in the 1850s), but because the tubers are often over two to three feet long (and just as deep), they are difficult for commercial harvest; New Zealand spinach (*Tetragonia tetragonioides*), a plant that thrives on summer heat and is easy to care for, in addition to being identical in taste to the time-honored variety; and the rainbow potato mix that includes the 'Yellow Finn', 'Peruvian Blue', 'Green Mountain' (it's white), and 'Blossom' (it's pink), making a never-to-be-forgotten potato salad.

Every page is decorated with pen-and-ink drawings of the various plants, a number of visuals plus dozens of hints on how to grow these garden treasures, as well as scores of delicious recipes to help the gardener in preparation.

But what, you may ask, is the value of heirloom seeds? "That's easy," said Jan.

> *An heirloom seed is a direct connection with the past. I know, for example, that a kale cultivar called 'Ragged Jack' has been around in gardens for over one hundred years. And before that it was probably growing for about a thousand years. That connection with the past is terribly exciting to me. And 'Ragged Jack' is now used in trendy restaurants and because it's valued today, most likely it will stick around for future gardeners. And besides they're unusually great food!*

JERRY BLACK OF OREGON EXOTICS RARE FRUIT NURSERY

Oregon Exotics Rare Fruit Nursery is stationed in Grants Pass, Oregon. Its one-hundred-page catalog is a fascinating mix of strange vegetables, ranging from the snake melon (*Trichosanthes anguina*) to the ichandarins—hardy members of the citrus family long cultivated in China and Japan—to the maple-flavored Arizona queen of the night cactus (*Peniocereus greggii*). To me, the most interesting part of the catalog deals with seed offerings that feature Chinese medicinal herbs, edible plants from China, and seeds from the Himalayas. In the new catalog for 1995, seeds of rare fruits, nuts, vegetables, and medicinal plants from the jungles and gardens of South America were introduced. It's a fascinating mix. As Jerry attests:

> *When I started farming it was on a little worm farm my dad was running near San Diego; the mid-1970s was a composting era, and worms were in vogue. The future looked bright, the weather was pleasant, and we had enough tax deductible horse manure to make anything grow.*

During that time, I started making trips to Mexico, importing blankets, huge Mexican papayas, and mangos—things that were not available in the United States. I'd bring back various fruits and sow the seeds in my dad's worm beds.

But by 1981, the worm business went bust and I left the farm. I borrowed my dad's beat-up pickup, a can of worms, and moved to Oregon in hopes of starting my own compost pile and again planting seeds of unusual fruits I might find. When plants survived the winter, I continued to experiment with early-ripening figs, kiwis, persimmons, and citrus varieties, especially after many local nurseries told me many of these fruits would not survive.

But Jerry had seen acres of kiwis growing in Canada, found thousands of fig trees growing eight hundred miles north of his home, so he knew there was a lot to be done. He started to travel, visiting top researchers in the United States, and went to the coldest regions of the earth, wherever the diversity of plants was the greatest. He went to Bolivia, Korea, Japan, Laos, Burma, and India, always looking for tropical fruits that would continue to do well in the Temperate Zone.

Although always frustrated with the Chinese government alone or together with my Chinese business partner, we formed a network of rural horticultural experts in some of the most forbidden Chinese mountains. We made three trips in search of a legendary fruit that supposedly remained hidden in an inaccessible high mountain area called the Northern God Peach. It's an evergreen vine, which bears softball-sized fruits, dripping with peachy juices. It survives at 6,000 feet where the snow piles five feet deep each winter. I've already found two of its relatives, both of them from cool, but milder forests. One is from the remote mountains of the Hani-Yao Autonomous region (along the border of northern Laos), and another species is from northern Burma, which has a beautiful green skin and yellow flesh, and is segmented like a giant raspberry. But the northern species was the most desirable because of its hardiness.

Two years ago, Jerry traveled through fifteen hundred miles of closed Chinese territory. He was joined by a cameraman from Hollywood who was scouting for a location to shoot a documentary. They entered areas that had never been explored and found tribesmen who had collected a close relative of the Northern God Peach, but had never found the real thing. Jerry returned to the states empty-handed.

Somewhere in China though, these fruits were ripening, then falling to the ground unseen in some forbidden mountain range. So he wired money to his Chinese partner and told him to find the fruit. Unfortunately, all they got that year was a blurry photograph from a villager.

But in the autumn of 1995, his Chinese partner set out alone and revisited the area they had once explored together. In an isolated forest, where parts of the population still live in caves, he found the plant and sent back photos and sixteen seeds.

"Imagine," said Jerry,

> *a broadleaf evergreen vine with commercial quality grapefruit-sized fruit the flavor of a peach. Now, with sixteen seeds of the hardier northern strain, the tribal name, and a big black dot on the map, we'll hopefully re-collect it next fall, and bring it to market.*

In South America, Jerry has looked for medicinal trees, attempting to set up a network of suppliers to help find seeds as they ripen. Recently, he went to Columbia to find one plant. He lived for weeks in the Andes and became friends with a farmer, a man who claimed he was the governor of the Incas, and his family. The plants they gave to Jerry had been passed down for generations.

Among the hundreds of plants that Jerry has collected, very few are known to the general public. According to Jerry:

> *There's a relative of the kumquat that could survive all the tough weather that comes down from Alaska. There's no name and it came from the high Australian desert. According to the collector, this fruit will take temperatures of at least 10° Fahenheit below zero and will get by on six inches of rain a year. Until I know better, I'm calling it a razlequat.*

ED RASMUSSEN OF THE FRAGRANT PATH

In 1982, Ed Rasmussen of Fort Calhoun, Nebraska, founded a mail-order seed company called The Fragrant Path. About his beginning he said:

> *When I started the business I realized that many catalogs in the trade paid little attention to a plant's scent, even those plants that were once grown primarily for their fragrance. Then we realized that many of the plants must be started from seed because plants were not to be had. But as the seedlings grew, and over the years began to flower and set seed themselves, we got a great reward for having persevered. There is a certain feeling for a plant that a gardener has started from that seemingly inanimate object called a seed.*

Fourteen years later, his catalog numbers sixty pages and includes such favorites as the original sweet pea (*Lathyrus odorata*), the incredibly fragrant mignonette (*Reseda odorata*), and for larger gardens, the sweet birch (*Betula lenta*).

The sweet pea was discovered in 1697 in Sicily by Father Cupani, who immediately sent seeds to a Dr. Uvedale in England. It was little more than a weedy plant but possessed a powerful and sweet fragrance. If it wasn't for that scent, it would surely have perished for it remained unimproved until 1870. But by 1900, there were some 264 varieties exhibited at London's Crystal Palace. Today we offer the old-fashioned or grandiflora varieties, not as large as some of the newer types, but all have the intoxicating perfume on which their fame is based.

As to mignonette, it's not a very ornamental plant but has long been grown for its distinctive and powerful fragrance still used in the compounding of perfume. It's often used in France for growing in pots on balconies and terraces, a fashion started by the Empress Josephine when Napoleon sent her seed during his Egyptian campaign. Mignonette is French for "little darling."

The sweet birch is native to the Eastern United States. While young, it demonstrates a pyramidal growth habit and has an outstanding dark cherrylike bark. All parts of the tree are used for the distillation of oil of wintergreen and when chewed, the twigs have a fine wintergreen flavor.

Running a seed business is the next best thing to the confusions found in the fashion industry. When Mr. Rasmussen writes catalog copy, it's usually just before Christmas, but he certainly knows how to start people thinking about spring. In the 1995 catalog for The Fragrant Path, he wrote,

I have just come out of the garden and there is a four-inch blanket of snow on the ground and fog in the air . . . besides it's nearly Christmas, and even though the shortest day of the year is at hand, there are still some citrus, jasmines, sweet olive, orchids, and a few other plants scenting the greenhouse air.

Back in the early 1970s, my brother and I moved to a 200 acre farm. There was nothing there except some big elm trees that were in the process of dying from Dutch Elm Disease, so we started a little nursery and grew many plants from seed. In those days there were big companies like Earl May but not too many of the little businesses.

When asked how he started in the seed business, he answered, "My nose," then added,

As kids, my brother Andrew, who today is a pathologist, and I were always planting things. We even tried to graft pears onto silver maples, so the love of plants was always there.

FRED SALEET OF THE BANANA TREE

There are two things to remember about Easton, Pennsylvania: First, there is an old railroad bridge that stands behind a McDonald's fast-food franchise where the stone abutments of that bridge are carpeted from top to bottom with ebony spleenworts (*Asplenium platyneuron*). The second claim to fame concerns The Banana Tree, Fred Saleet's company that salutes both rare seeds and rare plants, founded in Easton in 1960.

Fred Saleet specializes in a number of great tropical plants, including a wide range of heliconias (*Heliconia* spp.), plants known for their flowering bracts that may have the shape of a lobster's claw or a bird's beak. And for those of you who are continually searching for a plant that can, when blooming, impress the most sophisticated of gardeners, consider the Voodoo lily or *Amorphophallus riviera*. The Voodoo lily is a tropical plant that bears a notorious flower of great size and little beauty with a powerful, unpleasant odor because it is pollinated by flies.

> *I started the business back in 1960, probably because I've spent most of my life with tropical plants. After many years of traveling in the tropics, I've come to know Easton as the only shipping point and research center for work with researching tropical plants because much of the work is done in Costa Rica.*

The Banana Tree catalog's general listing of seeds is most impressive. The company deals with seeds for gingers, chocolate, true cinnamon, in fact thousands of different seeds, many not listed in the catalog. You can find, for example, over twenty-two species of Acacias (*Acacia* spp.), easily grown trees with spikes or globose heads of flowers that bear bright yellow or white balls of long stamens. Indigenous to Australia, they are commonly called wattles.

Said Mr. Saleet,

> *They do beautifully as indoor trees, but you have to provide bright light. Then after pruning them to size, they will flower. But remember that most of these trees have very hard seeds, that are best brought to germination by placing them in a cup of boiling water, then allowing them to soak for three days before planting.*

The species include the blue leaf wattle (*Acacia cyanophylla*), a shrubby, hedgelike tree that can reach a height of eighteen feet, has twelve-inch leaves, and bears yellow flowers, followed by five-inch-long pods that are constricted between the seeds.

The Banana Tree is also the only source I've ever found for the baobab tree (*Adansonia digitata*), also called the monkey-bread tree or the dead-rat tree. The tree's bark is used to make a fiber, and the young leaves are eaten as a vegetable, but its fame usually rests with the size of its huge trunk that grows to a thirty-foot width. The seedlings resemble a bottle when they are two years old.

Most gardeners are familiar with the Russian olive (*Elaeagnus angustifolia*), but The Banana Tree is a rare source for the seeds. Also on the unusual side is the Hawaiian woodrose (*Ipomoea tuberosa*), a tropical vine that can be grown indoors and produces yellow flowers, followed by a globular fruit in the form of a woody pod that looks exactly like a rose made of wood.

Said Mr. Saleet,

> *For those gardeners interested in things to eat, they should try* Jicama pachyrhyzus, *an increasingly popular vegetable that is used as a substitute for water chestnuts and eaten either raw or cooked. Easily grown in a mix of loamy soil with plenty of sand, they like a dry climate and can be grown indoors as a very attractive house-plant with lobed and heart-shaped leaves and light purple flowers. Remember, these plants must be kept warm and if grown outside, they produce a delicious underground tuber that can weigh in at five pounds.*

He paused for a moment.

"I have a love-hate relationship with the seed business. I want to love it, to savor it, but you get so worn out from the logisitics of running such a business that it often takes away from the fun."

As soon as he has time, though, he's off to work on next year's catalog!

ALAN D. BRADSHAW OF ALPLAINS

Alplains is short for Alpine Plants on the Plains. It's the name of a seed house located in Kiowa, Colorado, owned and operated by Alan D. Bradshaw, and dedicated to the preservation of the wildflowers found on the eastern plains of Colorado. Kiowa is about forty miles southeast of Denver, where necessary protections are needed during the winter.

"It's a challenge," said Mr. Bradshaw, "to grow anything in our severe climate." He continued,

> *Temperatures alone may fall to –30°F during the winter. In spring and summer almost constant winds howl around us, making water conservation a matter of top priority. We use a rock mulch that is really a natural scree, with pieces about a half-inch on the average. Then, depending on the part of the garden involved, I vary the mulch between quartz monzonite, to Pike's Peak granite that is a pinkish-tan to red, to a feldspar mix that is pinkish-white.*

There's also a ten-foot-high trellised fence on the north and west sides of the garden, not only to block the wind, but to catch snow where it's most needed as a protective mulch. The various rock gardens, plus a garden devoted to irises, cover about half an acre. But eventually, our gardens should occupy over two acres and allow us to grow a wide spectrum of floral diversity. An unheated, 500 square-foot greenhouse gives us the chance to grow the less hardy flowers from Australia, the Mediterranean region, South Africa, and the Mojave Desert.

*In our garden where hundreds of varieties are grown together, there is always the chance that unwanted hybridization may occur, especially within varieties like columbines (*Aquilegia *spp.) and the pinks (*Dianthus *spp.)*

In some instances, we pollinate by hand or rely on flowering time differences to protect the integrity of certain species. Seeds collected in the wild are even more likely to be genetically pure, although natural variation is common. But in the vast majority of cases, our seed will come true to name.

I asked about starting a seed business.

When I moved to Colorado, I met rock garden enthusiasts like Gwen Kelaidis, the editor of the Bulletin of the American Rock Garden Society, *and her husband, Paniotes Kelaidis of the Denver Botanic Garden. That led to a final link with native rock plants and the present business. As to dealing in seeds rather than plants, I can only say that seeds are a more convenient way to distribute germ plasm. They can be shipped around the world without any hassles. Just try that with plants!*

About the mysteries of stratification, Mr. Bradshaw is quick to point out that this is simply nature's way of holding up germination until growing conditions are more favorable.

Usually a cold spell of between 34°F and 40°F will remove the germination inhibitors, and the seed, knowing that winter is past, will soon feel it's safe to sprout. For most of our seeds, three to four weeks is sufficient time to simulate an entire winter. And if you feel that flats will take up too much refrigerator space, the seed is easily mixed with a little moist sowing medium (or even clean sand), put into a small plastic bag, then tucked in a corner, maybe next to the butter. But remember to avoid freezing.

The catalog is well documented and features plenty of information on seed pretreatment, including the necessity of light for germination, how deep to cover the seed, and whether or not the seed coat should be nicked. Germination codes include how much, if any, stratification is needed and whether the seed has been collected in the wild. Finally, Alplains provides the county and state locations, plus the approximate elevation in feet and meters. And if that is not enough, the approximate height and width of the mature plant is given in centimeters with a zone rating included. "Although," said Mr. Bradshaw, "this last item is an approximation at best, and always errs on the conservative side."

Various seed receptacles including (clockwise from right) money plant (Lunaria annua), *shoofly plant* (Nicandra phsalodes), *love-in-a-mist* (Nigella damascena), *bells-of-Ireland* (Moluccella laevis) *and the poppy* (Papaver 'Hungarian Blue').

CHAPTER 13

The Seed Nurseries, Seed Exchanges, and Some Great Sourcebooks

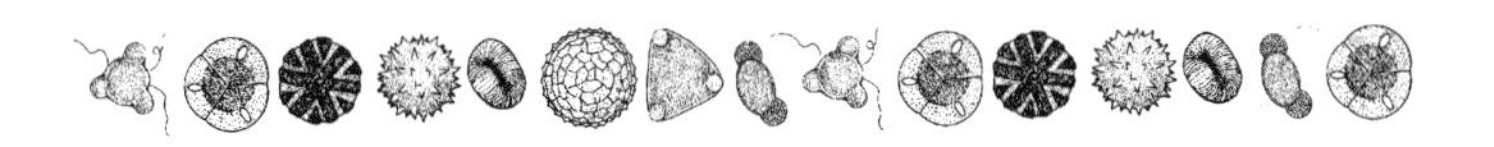

In the foreword of *A Thatched Roof* by Beverley Nichols (Jonathan Cape, London, 1949), Nichols talks about a gardener's house, a place where there is always a little mud on the floor, and:

> *In every corner there is evidence of the prevailing passion of its owner. The bottom of nearly every china ornament is black with seeds. There are also seeds in nearly all the cigarette boxes . . . seeds which stick to the cigarettes, and produce a strange smell of burning wool when lit, or even miniature explosions, so that my friends, after one or two experiments, give a sickly smile and say, "No thanks, really . . . I'd rather have one of my own."*

Obviously, cigarettes are no longer held in such high esteem, but nothing has changed about seeds. And if you are at all like me, there will be seeds in many places around your house or apartment. But unless you collect your own, these seeds must come from somewhere, and some of the best sources are seed nurseries and seed exchanges.

After phoning each company to make sure the telephone number and catalog price were correct, I now realize there are more seed houses in America than one gardener could ever imagine. In addition to all these marvelous growers, there is also a great reference section for seed collectors.

SEED NURSERIES

Abundant Life Seed Foundation
P.O. Box 772
1029 Lawrence Street
Port Townsand, WA 98368
(206) 385-5660

Abundant Life is a nonprofit organization that deals with open-pollinated and organically grown seeds, including vegetables, grains, native plants, and garden flowers. The catalog price is $2.00.

Allwood Bros.
Mill Nursery
Hassocks, W. Sussex,
England BN6 9NB
(0273) 84-4229

Allwood Bros. is world famous for its work with garden carnations and pinks (*Dianthus*) and has created some of the all-time great cultivars in this group of flowers. It also has seed of many primulas and wildflowers from South Africa. The catalog price is $1.00 (in U.S. dollars).

Alplains
32315 Pine Crest Court
Kiowa, CO 80117
(303) 621-2247

As of 1995, Alan Bradshaw has produced five catalogs dealing with alpine plants from the plains. His selection is dedicated to the preservation of wild flora on the eastern Great Plains of Colorado. Many beautiful plants are featured. The catalog price is $3.00.

Apothecary Rose Shed
P.O. Box 194
Route 160
Pattersonville, NY 12137
(518) 887-2035

Shawn Schultz and Susan Cooper have been in business since 1986 operating this small seed house that deals with about two hundred varieties of herbs, including everlastings. The catalog price is $2.00.

Jim & Jenny Archibald
Bryn Collen
Ffostrasol, Llandysul
Dyfed, Wales SA44 5SB

The Archibalds are intrepid seed collectors that roam the world. They not only gather seeds, but also publish a great newsletter about their wanderings. The catalog price is $3.00 (in U.S. dollars).

Arrowhead Alpines
P.O. Box 857
Fowlerville, MI 48836
(517) 223-3581

Bod & Brigitta Stewart feature an imposing selection of perennial wild and garden flowers, specializing in many that are termed *Alpine*, but remember, that word refers more to growing conditions than to the place of origin.

Ashwood Nurseries
Greensforge
Kingswinford, W. Mid.
England, DY6 0AE
(0384) 40-1996

The owners, John Massey and Phillip Baulk, specialize in seeds of the lewisias (*Lewisia* spp.), marvelous rock plants, with most of them originating in the Pacific Northwest of the United States. They also carry hellebores (*Helleborus* spp.) and cyclamen (*Cyclamen* spp.). The catalog price is $2.00 (in U.S. dollars).

B & T World Seeds
Whitnell House, Fiddington
Bridgewater, Somerset
England TA5 1JE
(0278) 73-3209

David Sleigh sells seed for thousands of tropical plants, including palms, bromeliads, flowering trees (some of them night-blooming), daturas, and enough brugmansias to start you in a breeding program. The list is, frankly, incredible. Their major catalog lists twenty-five thousand entries and sells for $20.00. Send two International Reply Coupons for the master list of the various seed lists.

The Banana Tree
715 Northhampton Street
Easton, PA 18042
(215) 253-9589

Fred Saleet has been running The Banana Tree for more years than I care to remember. And for all the tropical seeds (and plants) he offers, the firm is located in Pennsylvania, not far from the Delaware River. They offer an eclectic selection of seeds, varying from the gum arabic tree (*Acacia senegal*) to the Mexican cycad (*Dioon edule*) to the jujube or Chinese date (*Zizyphus jujuba*). The catalog price is $3.00.

Berton Seeds Company, Ltd.
151 Toryork Drive, Unit 20
Weston, ON
Canada M9L 1X9
(416) 745-5655

Berton Seeds specializes in all sorts of vegetables, including beans, lettuces, peppers, herbs, and some flowers. Most of their seeds are imported from Italy. Their color catalog price is $12.00, but they will send you a free seed list.

Birch Farm Nursery
Gravetye
East Grinstead
England RH19 4LE
(0343) 81-0236

Birch Farm is the home of Will Ingwersen, a master of the rock garden and author of *Manual of Alpine Plants*. It's one of the best-known alpine nurseries in England. Thanks to the magic of credit cards, it's easy to buy seeds. The list is two International Reply Coupons (IRC), and the catalog price is $2.00 (in U.S. dollars), or 5 IRC.

Bluestem Prairie Nursery
Route 2, Box 106
Hillsboro, IL 62049
(217) 532-6344

Ken Schaal specializes in plants and seeds of the prairie plants of the Midwest, especially Illinois, including many great grasses. The catalog is free.

Boothe Hill Wildflower Seeds
23 B Boothe Hill
Chapel Hill, NC 27514
(919) 967-4091

Nancy Eaterling sells about thirty varieties of wildflower seeds, all native to the Southeast, organically grown, and field collected. The catalog price is $1.00.

Bountiful Gardens
18001 Shafer Ranch Road
Willits, CA 95490
(707) 495-6410

Bill and Betsy Bruneau manage this organically oriented seed business that not only sells its own open-pollinated flower, herb, and vegetable seeds, but also is the American representative for Chase Seeds of England, the company that merged with the Henry Doubleday Foundation after Mr. Doubleday's recent passing. The catalog is free.

W. Atlee Burpee Company
300 Park Avenue
Warminster, PA 18974
(800) 888-1447

Burpee has been gracing American gardens since 1876, selling hundreds of varieties of vegetable and flower seeds. They also sell mixes of various seeds for specialized gardens. The catalog is free.

D. V. Burrell Seed Growers Co.
P.O. Box 150
Rocky Ford, CO 81067
(719) 254-3318

The Burrells operate this seed house that is more than one hundred years old. They carry a wide range of seeds, including vegetables, melons, peppers, and a large selection of annual flowers. The catalog is free.

Bushland Flora
17 Trotman Crescent
Yanchep, WA
Australia 6035
(09) 561-1636

For those interested in striking out into new territory, nothing beats starting a collection of Australian plants. Bushland Flora carries a huge selection, including many mahy helichrysums or everlastings. The catalog price is $2.00 (in U.S. dollars) or three International Reply Coupons.

Callahan Seeds
6045 Foley Lane
Central Point, OR 97502
(503) 855-1164

When I called Frank Callahan to check on his phone number, I immediately became involved in a discussion of the beautiful *Calochortus nitidus* and its hardiness. Mr. Callahan sells about six hundred species of native Northwestern trees and shrubs, either in small packets or in bulk. The catalog price is $1.00.

Chadwell Himalayan Seed
81 Parlaunt Road
Slough, Berks.
England SL3 8BE
(0753) 54-2823

Chris Chadwell is another intrepid plant explorer who combs the mountains of the world for unusual seeds, selling shares in the enterprise (see page 165). He also has a catalog of various seeds, including many suited for beginning gardeners. The catalog price is $3.00 (in U.S. dollars).

Chehalis Rare Plant Nursery
2568 Jackson Highway
Chehalis, WA 98532
(206) 748-7627
Herbert Dickson deals only with primula seed by mail, featuring single, double, and show blooms. Write for the seed list, enclosing a business SASE.

Chiltern Seeds
Bortree Stile
Ulverston, Cumbria
England LA12 7PB
(01229) 581137
Chilterns publishes one of those catalogs that is easy to pick up but terribly hard to put down. With more than 270 pages, the number of seeds they carry is huge, and the wonderful history that's provided for each one is very, very entertaining. And remember, aside from Chilterns, Ulverston is also famous as being the birthplace of Stan Laurel. The catalog price is $4.00 (in U.S. dollars).

Christa's Cactus
529 W. Pima
Coolidge, AZ 85228
(520) 723-4185
Christa Roberts deals in seeds of cacti and succulents, including some of the caudiciforms that are so great as specimen plants. The catalog price is $1.00.

Companion Plants
7247 N. Coolville Ridge Road
Athens, OH 45701
(614) 592-4643
Peter and Susan Borchard sell seeds for more than two hundred varieties of plants, including perennials, annuals, some everlastings, and even a few woodland plants and wildflowers. The catalog price is $3.00.

Comstock, Ferre & Co.
P.O. Box 125
263 Main Street
Wethersfield, CT 06109
(203) 571-6590
An old American firm, Comstock, Ferre & Co., is now in its 175th year. They carry more than four hundred varieties of vegetable, herb, and flower seeds, both annual and perennial. The catalog price is $3.00.

Comstock Seed
8520 W. 4th Street
Reno, NV 89523
(702) 746-3681
Ed & Linda Kleiner harvest seed from all over the Southwest and offer a good selection of native plants and native grasses, especially those of the Great Basin region. They will also try to locate seeds for special requests. The catalog is free.

The Cook's Garden
P.O. Box 535
Moffits Bridge
Londenderry, VT 05148
(802) 824-3400
Back in 1977, Shepherd and Ellen Ogden created The Cook's Garden when it was a new and surprising thing to do (and people who grew vegetables the organic way also talked to space aliens). Well, the rest of the world caught up, and

now gardeners can have a great selection of vegetables, herbs, and flowers. The catalog price is $1.00.

William Dam Seeds
P.O. Box 8400
279 Highway 8, West
Flamborough
Dundas, ON
Canada L9H 6M1

In business for more than forty years, William Dam carries a number of vegetable seeds, many annual flowers, and a number of houseplant seeds. He is one of the few sources left for the castor bean (*Ricinus*). The catalog price is $2.00.

DeGiorgi Seed Company
6011 N Street
Omaha, NE 68117
(402) 731-3901

An old and well-established seed house from 1905, DeGiorgi Seed Company is now under new ownership but still offers a large selection of seeds for annuals, perennials, vegetables, and a number of ornamental grasses. The catalog price is $2.00.

Desert Enterprises
P.O. Box 23
Morristown, AZ 85342
(602) 388-2448

The proprietor, Judith Clement runs a wholesale enterprise, dealing with wildflower seed only by the pound, especially for revegetation and erosion control. I list it for the odd problem that requires so much seed. Price list is $1.00.

Desertland Nursery
P.O. Box 26126
11306 Gateway East
El Paso, TX 79926
(915) 858-1130

Desertland specializes in rare and hard-to-find seeds of Mexican and South American cacti, including some habitat-collected seeds for succulents and tropical trees. The catalog price is $1.00.

Desert Moon Nursery
P.O. Box 600
Veqiota, NM 87062
(505) 864-0614

Theodore and Candace Hodoba started with a plant nursery specializing in Southwestern native plants, but now they're in the second year of carrying seeds for thirty-nine plants from the high desert. The catalog price, including seed list, is $1.00.

Euroseeds
Mojmir Pavelka
P.O. Box 95, 74101 Novy Jicin
Czech Republic

Euroseeds supplies seeds collected from Africa, Spain, and the Alps, and includes many rare alpines. Send $2.00 (in U.S. dollars only) for listing.

Henry Field Seed &
Nursery Company
415 North Burnett
Shenandoah, IA 51602
(605) 665-9391

This Midwest nursery has been around since 1892 and continues to carry a grab bag of plants, including many vegetable seeds sold by packet or pound, plus a good selection of annual and perennial flower seed. The catalog is free.

Flowery Branch Seed Company
P.O. Box 1330
Flowery Branch, GA 30542
(404) 536-8380

Flowery Branch Seed Company offers a large selection of herbs and everlastings and a good selection of garden annuals and perennials. The catalog price is $3.00.

The Fragrant Path
P.O. Box 328
Fort Calhoun, NE 68023

Ed Rasmussen sells seeds of fragrant, rare, and old-fashioned plants. There are annuals, perennials, vines, prairie flowers, and even a section on salad vegetables that includes a bit of history with each one. He also sells plants for evening fragrance. Ed won't take phone calls. The catalog is $2.00.

Gleckers Seedsman
Metamora, OH 43540
(419) 923-5463

Gleckers has been in business since 1947 and still carries an unusual selection of seeds. Choices vary between the sprayless tomato (Glecker introduced it in 1951) to their mammoth Hungarian giant squash (often weighing one hundred pounds) and a few flowers, including a marigold called "Mediterranian Moon" imported from New Zealand. The catalog is free.

L. S. A. Goodwins & Sons
Goodwins Road
Bagdad So.
Tasmania, Australia 7030
(002) 68-6233

Goodwins specializes in seeds of Australian plants and growing its seeds is the next best thing to climbing Ayer's Rock. Goodwins also features a good selection of cyclamen (*Cyclamen* spp.) seeds. The catalog price is $1.00 (in U.S. dollars).

Green Horizons
218 Quinlan, #571
Kerrville, TX 78028
(210) 257-5141

The owner, Sherry Miller, told me that Green Horizons specializes in native plants, Texas wildflower seeds, and some herbs. The catalog is free, but send them a business-sized SASE.

GreenLady Gardens
1415 Eucalyptus Drive
San Francisco, CA 94132
(415) 753-3332
Anthony J. Skittone

Known for years under the name Anthony J. Skittone, this bulb nursery was the place to go for all the unusual flowering bulbs of South Africa. As part of their service, they offer many seeds of selected species of imported bulbs. The catalog price is $3.00.

Gurney's Seed & Nursery Company
110 Capital Street
Yankton, SD 57079
(605) 665-1930
Gurney's has been in business for 127 years and still publishes an old-fashioned catalog with lots of color pictures and all sorts of annual and perennial flower and vegetable seeds. The catalog is free.

Harris Seeds
60 Saginaw Drive
Rochester, NY 14623
(716) 442-0410
Harris Seeds is another one of those time-honored American seed companies that features a large selection of vegetable and flower seeds. The catalog is free.

H. G. Hastings
P.O. Box 115535
2350 Cheshire Bridge Road
Atlanta, GA 30310
(404) 755-6580
A horticultural legend in the Southeast, Hastings carries a wide choice of seeds, including vegetables, annuals, perennials, and wildflowers that are specifically suited to the southern climate, especially around Atlanta. The catalog is free.

Henrietta's Nursery
1345 N. Brawley Avenue
Fresno, CA 93722
(209) 275-2166
Jerry and Sylvia Hardaway carry a large selection of seeds, specifically of cacti and succulents. The catalog price is $1.00.

Holland Wildflower Farm
290 O'Neil Lane
Elkins, AR 72727
(501) 643-2622
Holland sells seed mixes and seed for individual species. A thirty-two-page illustrated guide to wildflowers is $3.25. The price list for the seeds is free.

Homan Brothers Seed
P.O. Box 337
Glendale, AZ 85311
(602) 244-1650
Nathan Young specializes in habitat-collected seed from native plants of the Sonoran and Mojave Deserts. The collection includes grasses, wildflowers, and shrubs. The catalog price is $1.00.

J. L. Hudson, Seedsman
P.O. Box 1059
Redwood City, CA 94064
Hudson publishes another one of those catalogs that is better reading than most contempory novels. He is known as an ethnobotanist; his knowledge about seeds is great and the selection is grand. The catalog price is $1.00.

The Thomas Jefferson Center for Historic Plants
Monticello
P.O. Box 316
Charlottesville, VA 22902
fax: (804) 977-6140
The center offers a broadsheet of historic vegetable and flower seeds usually grown at Monticello and many that were originally collected (and imported) by Jefferson himself. The catalog is free.

Jelitto
P.O. Box 1264
Am Toggraben 3
D3033 Schwarmstedt, Germany
Telephone: 01149-5071-4085

Jelitto's fascinating catalog includes hundreds of rock garden alpine plants, plus unusual annuals and perennials, including many new cultivars. They list more than two thousand varieties. The text is German, but the cultural instructions are in English. The catalog price is $5.00.

Johnny's Selected Seeds
Foss Hill Road
Albion, ME 04910

Johnny's offers many vegetable and flower seeds, some of them unique. A well-produced catalog is free.

J. W. Jung Seed Co.
335 S. High Street
Randolf, WI 53957
fax: (414) 326-5769

In business since 1907, this venerable institution continues to offer a broad selection of vegetable and flower seeds in an old-fashioned catalog with charming color illustrations. The catalog is free.

Karmic Exotix Nursery
Box 146
Shelburne, ON
Canada L0N 1S0

The nursery offers garden-collected seed from the wild, including a vast number of campanula (*Campanula* spp.) and all types of other rare and unusual rock garden plants. The catalog price is $2.00.

Kester's Wild Game Food Nurseries
P.O. Box 516
4582 Highway 116 East
Omro, WI 54963
(414) 685-2929

Kester's was the seed source when we stocked a one-and-one-half-acre pond in Cochecton, New York. Its seed selections are extensive, its catalog is descriptive, and it deals with not only water plants for erosion protection and beauty, but also plants that produce wildlife food sources. The catalog price is $3.00.

Kingfisher, Inc.
P.O. Box 75
Wexford, PA 15090
(412) 935-8255

Elizabeth Bair deals mostly with herb seeds; her catalog accurately describes each plant, giving good advice on germination. The catalog price is $1.00.

Kitazawa Seed Company
1111 Chapman Street
San Jose, CA 95126

This company offers oriental vegetable seeds with enough selections to enable the average gardener to explore an entire new world. The catalog is free.

D. Landreth Seed Company
P.O. Box 6426
180-188 W. Ostend Street
Baltimore, MD 21230
(410) 727-3922

Opening in 1780, Landreth is the oldest seed house in America, boasting Washington and Jefferson as customers. It offers a wide selection of old vegetable varieties, plus herbs and old-fashioned garden annuals. The catalog price is $2.00.

Liberty Seed Company
P.O. Box 806
128 1st Drive S.E.
New Philadelphia, OH 44663
(216) 364-1611

Liberty offers heirloom annuals, perennials, and vegetables, mostly open-pollinated, and most are well adapted to the Midwest. The catalog is free.

Little Valley Farm
5693 Snead Creek Road
Spring Green, WI 53588
(608) 935-3324

Call for seed list.

Mellinger's, Inc.
2310 West South Range Road
North Lima, OH 44452
(216) 549-9861

Mellinger's is a venerable American institution of plants, equipment, including bonsai pots, and a strangely varied selection of seeds that cannot be catagorized. The catalog is free.

Mesa Garden
P.O. Box 72
Belen, NM 87121
(505) 864-3131

Mesa offers seeds of unusual rock garden plants, including a number of fameflowers (*Talinum* spp.) more-or-less succulent herbs mostly native to N. America. Send $1.00 or three $0.32 stamps for an extensive seed listing.

Moon Mountain Wildflowers
P.O. Box 725
Carpinteria, CA 93014
(805) 684-2565

Becky Schaff of Moon Mountain carries about sixty species of wildflowers and has prepared ten or so mixes, some specifically prepared for certain states and others listed as general mixes, good just about anywhere. The catalog contains plenty of clear illustrations and all seeds are tagged as to the best climate areas. The catalog price is $3.00.

Native American Seed
3400 Long Prairie Road
Flower Mound, TX 75028
(214) 539-0534

Jan & Bill Neiman work with Texas wildflowers and native grasses, including buffalo grass (*Buchloe dactyloides*), a sod-forming grass that can live on as little as twelve inches of water a year and is well suited to areas where summers are hot and dry. Because they want other Texans to know about these wild beauties, the Neimans offer all Texas residents a free seed list, but for those out-of-staters who want to try some of these great plants, the list is $1.00.

Native Gardens
5737 Fisher Lane
Greenback, TN 37742
(615) 856-0220

Meredith and Ed Clebsch own and operate Native Gardens, growing native plants and selling packet seed for about one hundred species of wildflowers. Their catalog gives concise information on needed growing conditions, season of bloom, flower color, and habitat. The catalog price is $2.00.

Native Seeds, Inc.
14590 Triadelphia Mill Road
Dayton, MD 21036
(301) 596-9818
Dr. James Saunders offers individual packets of wildflower seeds, plus mixes in bulk. In business since 1979, the firm has very knowledgeable employees, and the plants in the catalog are well described. The catalog is free.

Nichols Garden Nursery
1190 N. Pacific Highway
Albany, OR 97321
(503) 928-9280
After forty-five years, Nichols still maintains its high standards of quality and a continually expanding (and exciting) collection of garden seeds, including many vegetables and a fine selection of flowers. The catalog is free.

Northplan/Mountain Seed
P.O. Box 9107
Moscow, ID 83843
(208) 882-8040
Loring Jones has been operating Northplan for twenty years. He specializes in seeds of native trees and shrubs, wildflower mixes, and a number of native ornamental grasses. The seed list is $1.00.

Oregon Exotics Rare Fruit Nursery
1065 Messinger Road
Grants Pass, OR 97527
(503) 846-7578
Reading Oregon Exotics' catalog is like traveling along the Himalaya Hump with owner Jerry Black, in search of the rare and unusual, but never leaving home. The catalog price is $2.00.

Palms for Tropical Landscaping
6600 S.W. 45th Street
Miami, FL 33155
(305) 666-1457
The owner, Carol Graff, has seeds available for many species of palms, but the varieties offered are never constant. The list is free, but please include a business-sized SASE.

Park Seed Company, Inc.
P.O. Box 46
Highway 254 North
Greenwood, SC 29648
Driving to their annual seed trials in June, you can tell you're getting close to Parks because of the different flowers blooming along the roadside that came from seeds that escaped from their landscaped acres. Parks offers a great selection of vegetables and flowers, including heirloom seeds. The catalog is free.

Theodore Payne Foundation
10459 Tuxford Street
Sun Valley, CA 91352
(818) 768-1802
The foundation is a nonprofit institution that honors Theodore Payne and all the marvelous wildflowers of California. It specializes in plants that both attract and feed wildlife, and the choices number in the hundreds. The seed catalog price is $2.00.

Peter Pauls Nurseries
4665 Chapin Road
Canandaigua, NY 14424
(716) 394-7397

This nursery is the headquarters for seed of carnivorous plants. Remember, there are many more species than just the Venus flytrap. The catalog is free.

Pinetree Garden Seeds
Box 300
New Gloucester, ME 04260
(207) 926-3400

A small, but thick catalog from Pinetree includes a very personal selection of vegetable and flower seeds. The catalog is free.

Plants of the Southwest
Agua Fria Route 6, Box 11A
Sante Fe, NM 87505
(505) 471-2212

A labor-of-love catalog from this company includes plenty of color illustrations of marvelous desert plants and information about sowing the seed. The catalog is $3.50.

Prairie Moon Nursery
Route 3, Box 163
Winona, MN 55987
(507) 452-1362

The owner, Alan Wade, specializes in wildflowers and grasses for both creating and restoring prairie gardens. The catalog price is $2.00.

Prairie Nursery
P.O. Box 306
Westfield, WI 53964
(608) 296-3679

This company is one of the original prairie seed companies. Its informative catalog contains color wildflower photographs, great descriptions, and plenty of potentially ornamental grasses. The catalog price is $3.00.

Putney Nursery, Inc.
P.O. Box 265
Route 5
Putney, VT 05346

Putney Nursery has been in business since 1929, selling wildflower seeds specifically suited for various environments. The catalog is free.

Redwood City Seed Co.
P.O. Box 361
Redwood City, CA 94064
(415) 325-7333

The company offers old-fashioned, open-pollinated seeds, mostly popular before the turn of the last century. The catalog price is $1.00.

Clyde Robin Seed Company
P.O. Box 2366
Castro Valley, CA 94546
(510) 785-0425

Here's the place to buy wildflower seeds—not only by the packet, but also by the pound. The catalog price is $2.00.

Rocknoll Nursery
7812 Mad River Road
Hillsboro, OH 45133
(513) 393-5545

In addition to a number of rock garden plants, Jan and Jerry Hopkins have separate seed lists of unusual plants for alpine gardens. The catalog price is $1.00.

Rocky Mountain Rare Plants
P.O. Box 200483
Denver, CO 80220
Rocky Mountain Rare Plants is a small seed house specializing in alpine plants native to the Rocky Mountains and to similar environments around the world. The owner, Gwen Kelaidis, is the editor of the *American Rock Garden Society Bulletin* and is a most knowledgeable seedsperson.

Roswell Seed Co.
P.O. Box 725
115-117 South Main Street
Roswell, NM 88202
(505) 622-7701
The owners, Walter Roswell and Jim Gill, specialize in vegetables and grasses for Arizona, New Mexico, Oklahoma, and Utah. The catalog is free.

Rust-En-Verde Nursery
P.O. Box 753
Brackenfell, South Africa 7560
(27 21) 981-4515
A. M. Horstmann
These seeds (and bulbs) that come from the Cape Floral Kingdom include some of the most beautiful flowers in the horticultural world. The catalog price is $1.00 (in U.S. dollars).

Sandeman Seeds
The Croft
Sutton, Pulborough
West Sussex
England RH20 1PL
7987-315
Carl Sandeman of Sandeman Seeds specializes in rare and unusual plants, including trees and shrubs, subtropical plants and palms, plus ornamental grasses and perennials. At one time he dealt only in wholesale, but now his lists are available to the public. Telephone for catalog.

F. W. Schumacher Co., Inc.
36 Spring Hill Road
Sandwich, MA 02563
(508) 888-0659
Schumacher Co. specializes in seeds of conifers, rhododendrons, and azaleas, but many others are listed in the free catalog.

Seederama
P.O. Box 3
Charleston, N.S.W. 2290
Australia
(049) 49 8195
More than twelve hundred varieties of Australian and imported seeds from dwarf palms (*Didyosperma caudata*) to clivea hybrids (*Clivia* spp.) are offered by this impressive seed house from the land down under. The packets tell you how to mix the soil, whether the seed needs to be soaked, how sowing is best achieved, final plant position, type of soil needed, and even when to fertilize. Seeds sell for $1.10 per packet. The catalog is free, but include $2.00 (in U.S. dollars) for air mail postage.

Seeds Blüm
Idaho City Stage
Boise, ID 83706
fax: (208) 338-5658
Seeds Blüm (a one-person company) offers a catalog that is full

of the sheer joy of growing plants from seed, with plenty of observations on life along the way. The catalog price is $3.00.

The Seed Guild
P.O. Box 8951
Lanark, Scotland
ML11 9JG

The Seed Guild is a newly formed Scottish company that acts as a clearinghouse for surplus seed from various botanic gardens around the world. Send $3.00 for details and a sample packet of seed.

Seeds of Change
P.O. Box 15700
Santa Fe, NM 87506
(505) 438-8080

Seeds of Change offers a selection of hundreds of varieties of vegetables, herbs, and flowers, all grown by organic methods and all open-pollinated, following the belief that F_1 hybrid seeds bring evolution to a standstill. The catalog price is $3.00.

Seeds Trust
High Altitude Gardens
P.O. Box 1048
308 S. River
Hailey, ID 83333
(208) 788-4363

Bill McDorman's philosophy is dedicated to sustainable agriculture. He encourages this by demonstrating the importance of saving seed and selling almost exclusively open-pollinated varieties of seed collected from wildflowers in an environmentally sensitive manner. They feature vegetable and herb seeds as well as wildflower and native grass seeds. It also features garden portfolios, or seed collections, with a theme, ranging from edible flowers to medicinal herbs. The catalog price is $3.00.

SEPI
Seeds & Plants International
B P 5, l'Emeraude, l'Ange
Gardien
Quebec J8L 2W7
(819) 281-8611
Canada

Majella Larochelle continues to collect unusual plants, including many alpine species. He also carries a number of ornamental grasses, herbs, and plants native to Canada. Lists are $2.00.

Sharp Brothers Seed
396 SW Davis St., Ladue
Clinton, MO 64735
(816) 885-7551

Call for the twenty-page catalog price on various wildflower and grass seeds.

Sharp Brothers Seed Company
P.O. Box 140
Healy, KS 67850
(316) 398-2231

Sharp deals in native grasses and wildflowers for the great Midwest. It also sells a number of lawn grasses, including the top-rated buffalo grass. The catalog price is $1.00.

Shepherd's Garden Seeds
6116 Highway 9
Felton, CA 95018
(408) 482-6910

Renee Shepherd was one of the first people to deal with European vegetables, including salad fixings and edible flowers. The catalog, priced at $1.00, is a delight to read.

R. H. Shumway Seedsman
P.O. Box 1
Route 1, Whaley Pond Road
Graniteville, SC 29829
(803)663-9771

Shumway Seedsman is another of the grand and venerable seed institutions that dot American history. Its catalog is illustrated with old-fashioned line art, but it carries an up-to-date selection of vegetables and flowers. The catalog is free.

Southern Exposure Seed Exchange
P.O. Box 158
North Garden, VA 22959
(804) 973-4703

The company is not really an exchange, but one that offers more than 450 varieties of open-pollinated, heirloom, and traditional vegetables, sunflowers, flowers, and herbs. Every year it introduces new varieties, but it also includes a number of great old-fashioned flowers, including cultivars of the evening primrose (*Oenothera* spp.), and for four o'clock fans (*Mirabilis jalapa*), the cultivar "Don Pedros," bearing yellow and magenta flowers. The catalog price is $3.00.

Southern Seeds
The Vicarage, Sheffield
Canterbury, New Zealand 8173
NZ 318-3814

Southern Seeds deals in the mountain plants of New Zealand, where hot, dry summers are the norm and winters see snow and below-freezing temperatures. Many plants offered must have perfect drainage to survive. The catalog price is $5.00 (in U.S. dollars).

Stock Seed Farms
28008 Mill Road
Murdock, NE 68407
(402) 867-3771

Call for the catalog on various prairie wildflowers, native plants, and grasses.

Stokes Seed Company
P.O. Box 548
Buffalo, NY 14240
(416) 688-4300

A Canadian company with roots in Buf-falo, Stokes has a large selection of vegetable and flower seed, and their free catalog includes a great deal of cultural information. I was born in Buffalo, and my mother's garden always had seeds from Stokes.

Sunrise Enterprises
P.O. Box 330058
West Hartford, CT 06133
(203) 666-8071

Sunrise has imported seed from China and Japan since 1976, offering an incredible selection of unusual vegetable and some flower seeds. The catalog price is $2.00.

Thompson & Morgan
P.O. Box 1308
Jackson, NJ 08527
(908) 363-2225

The first T & M catalog was issued in England in 1855, and they have been going like gangbusters ever since. The catalog is well-illustrated with many small, full-color pictures of various offerings, and the species covered are often mind-boggling. You must order early because with many of the rarer varieties, only small amounts of seed are available. The catalog is free.

Tomato Growers Supply Company
P.O. Box 2237
Fort Myers, FL 33902
(813) 768-1119

In business for over ten years, this mom-and-pop tomato business carries more than 240 varieties of tomato seed, plus a huge selection of peppers. New selections for 1995 include "Banana Legs," "Olé Hybrid," and "Tiffen Mennonite." If you can't find a tomato or pepper to suit your fancy here, better switch to brussels sprouts. The catalog is free.

Otis Twilley Seed Company
P.O. Box 65
Trevose, PA 19053
(800) 622-7333

Yet another old American seed house, Otis Twilley started in 1920. It offers a good selection of vegetable and flower seed. The catalog is free.

Vermont Bean Seed Company
Garden Lane
Fair Haven, VT 05743
(802) 273-3400

Vermont Bean Seed Company specializes in bean and vegetable seeds, plus a small selection of annual and perennial flower seeds. The catalog is free.

Vermont Wildflower Farm
Route 7
Charlotte, VT 05445
(802) 425-3500

In a Northeast garden, the Allens test wildflower seeds and seed mixes and have the knowledge to know which plants will do best in your particular area. They stock more than forty species of wildflowers, and the various mixes usually contain from twenty to twenty-five species. The catalog is free.

John Watson Seeds
24 Kingsway, Petts Wood
Orpington, Kent
England BR5 1PR
689-822494

The Watsons travel around collecting rare and unsual seeds. Their last trip was to Peru. They issue a seed list that is free, but send two International Reply Coupons.

Wild and Crazy Seed Company
P.O. Box 895
Durango, CO 81302
(303) 259-6385

A small company, Wild and Crazy deals with wildflower seeds of the Southwest. The catalog price is $2.00

Wild Seed
P.O. Box 27751
Tempe, AZ 85285
(602) 345-0669
This company deals in seeds of Southwestern wildflowers and native plants, including a number of native grasses. The catalog is free.

Wildseed Farms
P.O. Box 308
Eagle Lake, TX 77434
(409) 234-7353
Wildseed Farms offers wildflower seeds gathered from across the country, many included in regional seed mixes. The catalog is free.

Roy Young, Seeds
23 Westland Chase, West Winch
King's Lynn, Norfolk
England PE33 0QH
(0553) 840867
Roy and Sheila Young have specialized in seeds of cactus and succulent plants for more than forty-five years. Their seed list numbers hundreds of choices, beginning with *Acanthocalycium glaucum* and ending with *Zygophyllum microcarpum*. The catalog price is $2.00 (in U.S. dollars) or three International Reply Coupons.

SEED EXCHANGES CONNECTED WITH VARIOUS PLANT SOCIETIES

In addition to the commercial seed houses, there are hundreds of individual societies that are dedicated to distributing both information and actual seeds to their members, many throughout the world. Most of these societies publish informational and professional journals and bulletins advising of society events and the availability of seeds to the members.

Alpine Garden Society
AGS Centre, Avon Bank
Pershore, Worcestershire
England WR10 3JP
01386 554790
Membership brings you a stunning full-color *Quarterly Bulletin of the Alpine Garden Society* along with a yearly seed exchange with selections numbering in the thousands, including both the rare and the unusual. I await the list every year like a child waiting for Santa. Membership is about $26.00 a year.

American Conifer Society
Charlene Harris, Executive Secretary
827 Brooks Street
Ann Arbor, MI 48103
(313) 665-8171
Members receive the newsy *The American Conifer Society Bulletin* and a yearly seed exchange dealing with conifer species and cultivars. Membership is $25.00 a year.

American Calochortus Society
P.O. Box 1128
Berkeley, CA 94701
This society is devoted to some of the most beautiful wildflowers in the world. Membership includes participation in a seed exchange. Membership is $4.00 a year.

American Fern Society
W. Carl Taylor, Secretary
Botany Department
Milwaukee Public Museum
800 West Wells
Milwaukee, WI 53233
The Fiddlehead Forum has all types of articles on ferns. The more scholary *American Fern Journal* ($7.00) is for the more professional grower. Also offered is a spore exchange. Basic membership is $8.00 a year.

American Horticultural Society
7931 East Boulevard Drive
Alexandria, VA 22308
(800) 777-7931
Members receive six issues of a marvelous, full-color magazine, plus six news bulletins and a good seed exchange every spring. Membership is $35.00 a year.

American Penstemon Society
1569 Holland Court
Lakewood, CO 80232
(303) 986-8096
Penstemons are such beautiful American wildflowers that they derserve a society of their own. New members receive a very good *Manual for Beginners*, plus a great seed exchange. Membership is $10.00 a year.

American Rhododendron Society
P.O. Box 1380
Gloucester, VA 23061
(804) 693-4433
Members receive the scholarly, full-color publication *The American Rhododendron Society Bulletin* four times a year, plus entry to a great and comprehensive seed exchange. Selections are vast and include hundreds of choices, including open-pollinated and hand-pollinated varieties. Membership is $25.00 a year.

Arizona Native Plant Society
P.O. Box 41206
Tuscon, AZ 85717
Write for information about its magazine and the seed exchange.

The Hardy Plant Society
710 Hemlock Road
Media, PA 19063
(215) 566-0861
The Hardy Plant Society (HPS) started as a great English organization and now has chapters in the United States. Membership includes a good newsletter, plus entry into a fantastic and professional seed exchange that deals mainly with garden annuals and perennials. In 1994, the HPS was starting a Short Viability Seed Exchange. The object of such a scheme is to make available fresh, damp-packed seed of the species that are difficult or impossible to germinate from dry seed. Jeffersonia spp., *Salix* spp., *Trilium* spp., and *Helleborus* spp. come to mind. Membership is $12.00 a year.

The Herb Society of America
9019 Kirtland Chardon Road
Mentor, OH 44060
(216) 256-0514

Members receive publications about herbs and a seed exchange. Membership is $35.00 a year.

The Hoya Society International
Christine Burton
P.O. Box 1043
Porterdale, GA 30270

Membership entitles you to four issues of the scholarly *The Hoyan* and a seed exchange. Christine Burton, a knowledgeable hoya aficionado, does much of the society's work without volunteer help. This is the original hoya society in America. Membership is $20.00 a year.

International Carnivorous Plant Society
Fullerton Arboretum
California State University
Fullerton, CA 92634
(714) 773-2766

Members receive four issues of *The Carnivorous Plant Newsletter* and join a seed exchange. Membership is $15.00 a year.

Louisiana Native Plant Society
P.O. Box 393
Blanchard, LA 71009

Members receive four issues of the *Louisiana Native Plant Society Newsletter*, plus a seed exchange. Membership is $10.00 a year.

Minnesota Native Plant Society
220 BioScience
1445 Gortner Avenue
St. Paul, MN 55108

Membership includes a newsletter and a seed exchange. Write for information.

Mississippi Native Plant Society
P.O. Box 2151
Starkville, MS 39759
(601) 324-0430

Membership includes a newsletter and a seed exchange. Membership is $7.50 a year.

Missouri Native Plant Society
P.O. Box 20073
St. Louis, MO 63144
(314) 577-9522

Membership includes a newsletter and a seed exchange. Membership is $9.00 a year.

National Wildflower Research Center
4801 La Crosse Boulevard
Austin, TX 78739
(512) 292-4200

This new national center, devoted to wildflowers, publishes hundreds of fact sheets about wildflowers along with suggestions of places where native plants and seeds can be purchased. Membership includes a bimonthly newsletter. Membership begins at $25.00 a year.

Native Plant Society of Texas
P.O. Box 891
Georgetown, TX 78627
(512) 863-7794

Membership includes a newsletter and a seed exchange. Membership is $20.00 a year.

North American Lily Society
P.O. Box 272
Owatonna, MN 55060
(507) 451-2170
Membership includes a newsletter and a seed exchange. Membership is $12.50 a year.

North American Rock Garden Society
Executive Secretary
P.O. Box 67
Millwood, NY 10546
This is a notable organization with the great *Rock Garden Quarterly* that comes four times a year, plus a giant seed exchange for members. Membership is $25.00 a year.

Northern Nevada Native Plant Society
P.O. Box 8965
Reno, NV 89507
(702) 358-7759
Membership includes newsletter and a seed exchange. Membership is $7.50 a year.

Northwest Horticultural Society
Yoosun Park
V. Isaacson Hall, University of Washington, GF-15
Seattle, WA 98195
(206) 527-1794
Membership includes the *Pacific Horticulture* and a seed exchange. Membership is $35.00 a year.

Ohio Native Plant Society
6 Louise Drive
Chagrin Falls, OH 44022
(216) 338-6622
Membership includes a newsletter and a seed exchange. Membership varies according to chapter.

Passiflora Society International
3900 W. Sample Road
Coconut Creek, FL 33073
(305) 977-4434
Membership includes a newsletter and a seed exchange. Membership is $15.00 a year.

Peperomia & Exotic Plant Society
100 Neil Avenue
New Orleans, LA 70131
(504) 394-4146
Membership includes a newsletter and a seed exchange. Membership is $7.50 a year.

The Royal Horticultural Society
80 Vincent Square
London
England SW1P 2PE
01733 898100
Membership in the RHS brings you twelve issues of a great magazine called *The Garden* and entry into another one of those worldwide seed exchanges. At this writing membership is about $32.00 a year.

The Scottish Rock Garden Society
Mr. Ian Aitchison
20 Gorse Way
Formby, Merseyside
Scotland L37 1PB
0704 876382
Membership brings four issues of *The Rock Garden*, plus another one of those mouth-watering seed

exchanges with selections numbering in the thousands. Membership is about $25.00 a year.

The Sedum Society
10502 N. 135 W.
Sedgwick, KS 67135
(316) 796-0496
Membership includes a newsletter and a seed exchange. Membership is $17.50.

Seed Savers Exchange
3076 North Winn Road
Decorah, IA 52101
(319) 382-5990
Thousands of backyard gardeners are members of the Seed Savers Exchange, a marvelous organization located in Iowa. Under the direction of Kent Whealy, Seed Savers began as a small organization dedicated to saving seeds of vegetable varieties from extinction. Why? Because many family-owned seed companies continue to be sold to agrichemical companies or bow to the pressures of the "You've-got-to-have-a-new-model-vegetable-every-year" mentality. As the Garden Seed Inventory (3d edition) points out, "230 seed companies [in business] in 1984, were no longer in business by 1987." Membership includes a copy of the *Seed Savers Yearbook* (see page 163) and the knowledge that you are personally involved in saving the heritage of American seeds. For those gardeners too busy to call, a color brochure that details all the activites of the Seed Savers is available for $1.00, Write to SSE, Rte. 3, Box 239, Decorah, IA 52101. Membership is $25.00 a year.

Sino-Himalayan Plant Association
Chris Chadwell
81 Parlaunt Road
Slough, Berks.
England SL3 8BE
Membership includes a newsletter and a seed exchange, including seeds collected on various Sino-Himalayan trips. Write for information and include one IRC.

Species Iris Group of North America
150 North Main
Lombard, IL 60148
Membership includes a newsletter and a complete seed exchange. Write for information.

Western Horticultural Society
P.O. Box 60507
Palo Alto, CA 94306
Membership includes a subscription to *Pacific Horticulture* and a seed exchange. Membership is $25.00 a year.

EQUIPMENT AND SUPPLIES

The following firms carry complete selections of various seed-sowing supplies along with pots, flats, labels, and other, sometimes useful, sometimes odd, products for the home sower.

Charley's Greenhouse Supplies
1569 Memorial Highway
Mount Vernon, WA 98273
(800) 322-4707
Charley's has an amazing number of greenhouse supplies, including many products for growing plants and seedlings. The catalog is $2.00.

Gardener's Supply Company
128 Intervale Road
Burlington, VT 05401
(802) 863-1700
Gardener's Supply Company carries a wide choice of seed-sowing supplies, plus a number of lights and racks for growing seeds indoors. The catalog is free.

A. M. Leonard, Inc.
241 Fox Drive
P.O. Box 816
Piqua, OH 45356
(800) 543-8955
Leaonard's eighty-page catalog for the nursery industry sells nursery flats, jiffy pots by the thousands, and various heating systems. The catalog price is $1.00.

Mellinger's
2310 W. South Range Road
North Lima, OH 44452
(216) 549-9861
In addition to selling seeds, Mellinger's offers many items for the home nursery and seed sower. The catalog is free.

Walt Nicke
36 McLeod Lane
P.O. Box 433
Topsfield, MA 01983
(508) 887-3388
Walt Nicke was the first mail-order garden supply company that dealt with all types of propagating and seed-growing equipment, in addition to a number of English gardening tools. The catalog is free.

SOURCEBOOKS FOR SEED COLLECTORS

Finally, in the world of seeds, there are four books that are extremely useful to gardeners who are continually searching for new ways to expand their seed horizons.

Andersen Horticultural Library's Source List of Plants and Seeds

Occasionally, some institution takes on the sort of task best described as Augean, referring to cleaning out the Augean stables, the ignored-for-thirty-year-home of three thousand oxen. In the world of horticulture, the Andersen Horticultural Library of the Minnesota Landscape Arboretum of the University of Minnesota is to be congratulated for another difficult job of compiling their *Source List of Plants and Seeds.* The newest edition was compiled in 1993. The last issue was finished in 1987.

From more than 400 selected nursery catalogs in the United States and Canada, the editors gleaned thousands of pages to provide the sources for more than 3,000 rhododendrons, 2,000 tomatoes, daylilies, 250 hybrid tomatoes, 200 maple cultivars, 125 petunia hybrids, plus thousands of other listings for vines, perennials, annuals, bulbs, and vegetables. In total, more than 47,000 plants are listed.

How do they do it? Briefly, the criteria for the listings were the following: The plant's scientific name had to be more or less correct; the nurseries had to ship their plants; and the shipping had to reach a good geographic distribution. For wild plants, the Anderson Horticultural Library only lists nurseries that propagate their own plants and do not collect in the wild.

Using the book is simplicity itself. Look under *Pinus banksiana* 'Aunt Fidget', and you might find a string of numbers like 56,p5,88,p78,54432,p45. The codes 56, 88, and 54432 are code numbers for nursery catalogs; the number preceded by the letter *p* indicates the page number of the respective catalog where the plants can be found. Furthermore, any catalog with a number of 50,000 or more is a wholesale source. Thus we know the first two are retail, and the last nursery is wholesale. The nursery listings include the full address and telephone numbers and are cross-referenced in a state-by-state geographical breakdown.

If the plant is available in both seed and plant form, seed sources (or in the case of ferns and spores) are indicated by the letter *s*. For example, 49,p45,s,p51 indicates that both plants and seeds are available.

To make things even easier, there is a cross-reference of common names, so that by knowing baby's-breath, you can find its scientific name, *Gypsophila paniculata;* or by knowing blue barberry, you can find *Mahonia aquifolium.*

To be on the cutting edge of scientific names, the library uses *The New Royal Horticultural Society of Gardening, Hortus Third,* and other even more scientific sources like the *New Britton & Brown Illustrated Flora,* to make sure everything is as correct as possible.

The softcover is available for $34.95 from Andersen Horticultural Library, Minnesota Landscape Arboretum, 3675 Arboretum Drive, Box 39, Chanhassen, MN 55317. Why not ask your library to get a copy?

The Bernard E. Harkness Seedlist Handbook

The first *Seedlist Handbook* (The Timber Press, Portland, 1974) began as a sourcebook for alpine and rock gardeners who used the various seed exchanges of the American Rock Garden Society, the Alpine Garden Society, and the Scottish Rock Garden Society. The *Seedlist Handbook* first appeared in 1974, and again in 1976 and 1980, covering the seeds offered from 1971 to 1979. The present title is *The Bernard E. Harkness Seedlist Handbook* with the second edition published in 1993.

Why, you may ask, was such a book needed? Because these three seed exchanges deal with thousands of seeds, many rare, many from forgotten gardens, many newly collected, and many not referenced in the standard sources of horticultural knowledge like *Hortus Third*, *The Royal Horticultural Society Dictionary of Gardening*, or more regional books like the *Manual of the Vascular Flora of the Carolinas*. Entries include the name of the species or cultivar, both the habit and hardiness of the plant, plant height at maturity, ornamental or other distinguishing features, the country or area of origin, and references to more extensive descriptive, cultural, and landscaping information.

For example, if you were confronted with *Mutisia pulshella* (a group of about sixty climbing shrubs, native to South America) and you looked for the plant in *Hortus Third*, you would not find a reference. But in the *Seedlist Handbook*, on page 306, you will note that *M. pulshella* is really known as *M. spinosa* var. *pulchella*, a plant with white leaves that are tomentose beneath, is a greenhouse plant, a climbing vine growing to twenty feet, bears pale rose flowers, and hails from Chile and Argentina.

After Mr. Harkness's death in 1980, Mabel G. Harkness, who majored in botany at the University of Rochester and later worked as a reference librarian, carried on the work, and she continues to update the information from the three seed societies.

It is published by Timber Press, Inc., 133 S.W. Second Avenue, Suite 450, Portland, OR 97204. Telephone: (503) 277-2878. The cost is $29.95 (softcover).

The Garden Seed Inventory

Published by the Seed Savers Exchange (SSE), the *Garden Seed Inventory* remains one of the great reference books for vegetable growers. Now in its third edition (with the fourth edition scheduled for release at the end of 1995), the present publication is a comprehensive inventory of 215 United States and Canadian mail-order seed catalogs. The book also contains descriptions of 5,291 standard vegetables and lists of the companies offering each variety.

The *Garden Seed Inventory* was developed as a preservation tool in an attempt to halt the continued destruction of hundreds of excellent vegetable varieties in danger of or actually being destroyed. The revenues from the sales of the *Garden Seed Inventory* are used to buy samples of endangered vegetable varieties while souces still exist and to develop SSE's Heritage Farm, where collections of rare food crops are maintained and are on display.

If you look up peppers (*Capsicum annuum*), you will find graphic evidence of the decline in open-pollinated varieties and descriptions of all the currently available varieties: for example, 'Albina', 'Big Red', and 'Bull Nose', and where to find them.

Books are available in softcover for $17.50 (prepaid postage) and $25.00 hardcover (prepaid postage) by writing to SSE, Rte. 3, Box 239, Decorah, Iowa 52101.

Collecting, Processing, and Germinating Seeds of Wildland Plants

Collecting, Processing and Germinating Seeds of Wildland Plants is written by James A. Young and Cheryl G. Young. Dr. Young is a research scientist for the U.S. Department of Agriculture and is considered an expert on native North American seeds. Cheryl Young, his wife and coauthor, is not only a knowledgeable gardener, but also the organizer of the literature used to document the text. The Youngs spend their summers raising roses and their winters in hybridizing African violets.

The tone of this book is serious. After a few chapters devoted to seeds, harvesting seeds, and seed cleaning, the Youngs take a number of pages to discuss the germination of tree seeds, including conifers and broadleaf evergreens, seeds of shrubs, a number of herbaceous species, and the germination of grasses. Annotations include hints such as: The cappato or tule-potato (*Sagittaria latifolia*) has seeds that require five to seven months of cold water storage before even moderate germination will occur, and the seeds of *Coreopsis lanceolata* need light for germination. I was surprised to learn that the germination of mistletoe seeds (*Arceuthobium* spp.) could be enhanced with a treatment including hydrogen peroxide.

The book is published by Timber Press, 133 S.W. Second Avenue, Suite 450, Portland, OR 97204. Telephone: (503) 227-2878.

Garden Flowers from Seed

Garden Flowers from Seed by Christopher Lloyd and Graham Rice (Viking, London, 1991), is a thoroughly wonderful book about raising garden plants from seed. The authors are both knowledgeable plantsmen. Christopher Lloyd is a respected writer of ten garden books and dozens of magazine articles. He is also the owner and chief gardener at Great Dixter, that most famous of English garden estates, located in the beautiful countryside of East Sussex. Graham Rice, the founder of the

Growing from Seed magazine, writes a number of columns for various English garden magazines and has authored three books on gardening.

The two men begin the text by discussing methods of seed sowing, followed by their favorite containers and composts. They then quickly turn to annotated entries on 219 pages of text in "The A to Z of Garden Plants." The entries run the gamut from hints on germinating acacias (*Acacia* spp.) to growing gas plants (*Dictamnus albus*) from seed and talking about the volatile oils produced by the plant that can be lit on hot summer nights with a match touched to a seedpod. Examples include sage advice about germinating henbane (*Hyoscayamus niger*), a very poisonous plant, recommending that seedlings not be transplanted (advice that I certainly will heed after the failures I've experienced with this particular plant).

The book is distributed by Timber Press, 133 S.W. Second Avenue, Suite 450, Portland, OR 97204. Telephone: (503) 227-2878.

Growing and Propagating Wild Flowers

Growing and Propagating Wild Flowers by Harry R. Phillips (The University of North Carolina Press, Chapel Hill, 1985) is a user-friendly book that deals with the propagation of hundreds of native wildflowers. It also includes notes on cultivation, uses in the garden or landscape, and related species. The home gardener will find useful sowing information on beard tongues (*Penstemon smallii*), including advice on sowing seeds immediately; butterfly weed or pleurisy root (*Asclepias tuberosa*), another seed that should be sowed immediately upon collection; and the knowledge that the seeds of our native passion flower or Maypops (*Passiflora incarnata*) may take two years for germination to occur, and then it's often poor and sporadic. There is even a section on carnivorous plants, including Venus'-flytrap (*Dionaea muscipula*), sundew (*Drosera* spp.), butterworts (*Pinguiculata* spp.), and the various pitcher plants (*Sarracenia* spp.).

The Reference Manual of Woody Plant Propagation

The complete title of this most comprehensive book on plant propagation is *The Reference Manual of Woody Plant Propagation, From Seed to Tissue Culture*, by Michael A. Dirr and Charles W. Heuser, Jr. It is described as a practical working guide to the propagation of more than eleven hundred species, varieties, and cultivars. The first chapter deals with seed propagation, and the encyclopedia section runs 143 pages, giving detailed information about raising plants from seed, running the gamut from the trident maple (*Acer buergeranum*) to Virginia sweetspire (*Itea virginica*) to the Wolf lilac (*Syringa wolfii*).

The book is published by Timber Press, 133 S.W. Second Avenue, cost Suite 450, Portland, OR 97204. Telephone: (503) 227-2878.

Seed Savers Yearbook

The *Seed Savers Yearbook* is published each January by the Seed Savers Exchange (SSE) and contains names and addresses of more than one thousand members of the SSE in addition to more than ten thousand listings of rare and unusual vegetables that are offered to members in this giant seed exchange. Seeds are obtained by writing directly to the members who are listed in the yearbook. Since SSE was founded in 1975, almost a half million seed samples that would otherwise be extinct have been distributed to members. It is only available to members of the SSE (see above).

Seed to Seed

Seed to Seed, by Suzanne Ashworth, is subtitled *Seed Saving Techniques for the Vegetable Gardener.* "For some years now," she writes, "I have been one of these seed saving gardeners [but] like many of my fellow seed savers, I have often been frustrated by not having access to detailed information on saving seeds from garden vegetables." So rather than waiting for somebody else to gather together the strings of research, Miss Ashworth has created a reference book about saving vegetable seeds, answering most questions that gardeners ask.

For example, if a gardener was interested in garbanzos, or chickpeas (*Cicer arietinum*), upon consulting the book, they would find a detailed description of the plant, including the fact that the leaves and stems are covered with glabular or sticky hairs. These hairs exude malic acid, which in India is collected for use as either vinegar or medicine. Readers would also learn that garbanzo flowers are self-pollinating, but because they will also cross-pollinate, the plants should be either caged or isolated by a half-mile to ensure seed purity. Information follows on collecting seeds, storing seeds, and learning that seeds maintain 60 percent germination for three years when stored in a cool, dark, dry location.

Books are available in softcover for $20.00 (prepaid postage) by writing to SSE, Rte. 3, Box 239, Decorah, Iowa 52101.

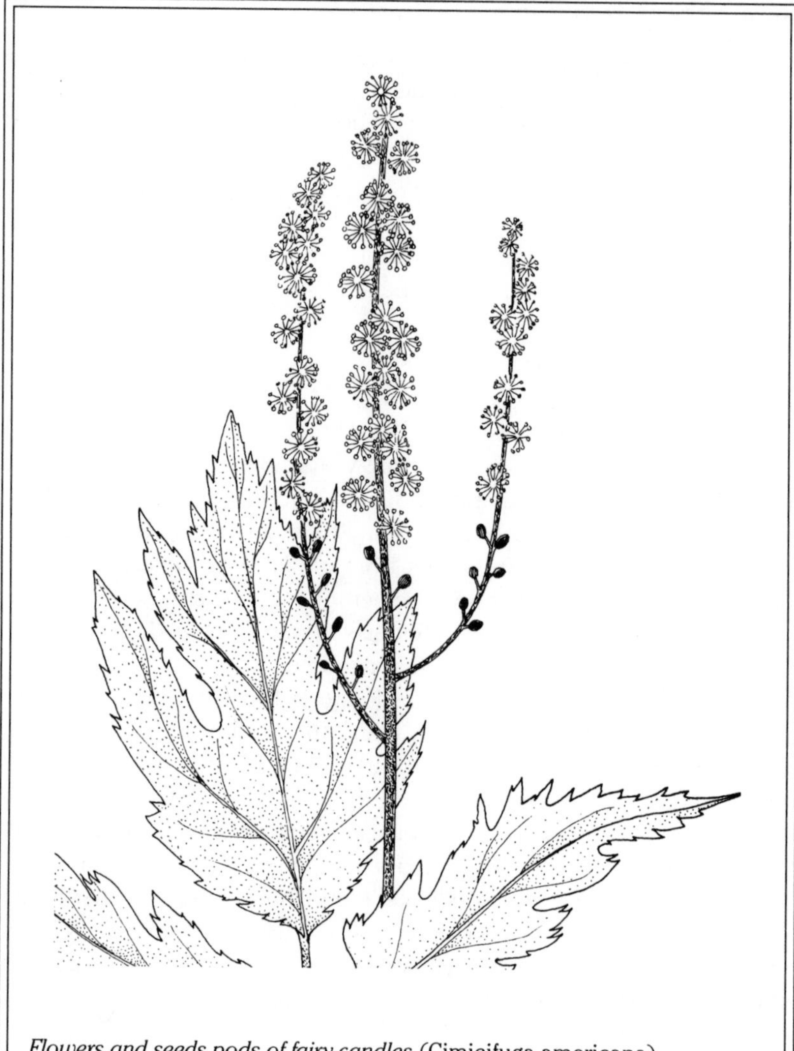

Flowers and seeds pods of fairy-candles (Cimicifuga americana).

CHAPTER 14

The Seed Collectors

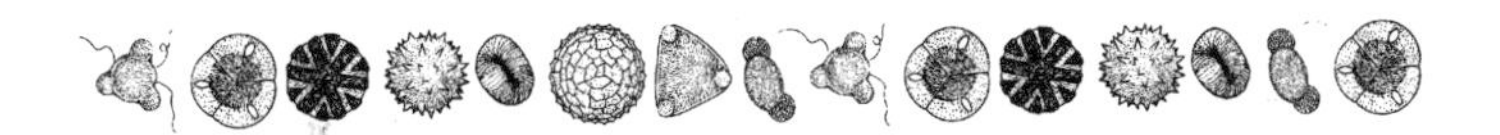

CHRIS CHADWELL AND THE SINO-HIMALAYAN PLANT ASSOCIATION

The history of horticulture is full of adventurers: men and women who are generally offended by grass except to know that it's always greener somewhere else and who will high-tail it there with all deliberate speed. Chris Chadwell is a contemporary plant explorer following in the footsteps of Frank Kingdon Ward, Ernest Wilson, and Reginald Farrer, except this thirty-six-year-old man doesn't collect plants, he collects seeds of the rare and the near-rare in cultivation.

The gardener who thinks of this adventure as a lark has probably never camped beyond his backyard or when out in the great beyond has returned to camp to enjoy good food and warm, clean clothes.

Not so with Mr. Chadwell. His expeditions are in the real world of terrible weather, limited medical care, boring but packable food, and a bed on the hard ground. He said,

> *My expeditions are a success if people come back the best of friends. Back in 1980, I joined my first expedition to the Western Himalaya Mountains and it could not have managed to be a more disagreeable time, with problems like the inevitable attacks of dysentery to a complete lack of team spirit. That first expedition was my expedition to hell.*
>
> *We made the same trip three years later and the worst that happened to me was narrowly surviving being swept away by a glacial river and somersaulting into a whirlpool. But in the end I think that botanical knowledge is less important than knowing you are part of a team and willing to spend some time working for the rest of the crew.*

Langley New Town in Berkshire, England, is Mr. Chadwell's base camp. Between expeditions, he and his wife, Dorothy, can be found there, where he operates a business selling Himalayan plant seed. The trial garden for his plants is a raised patch beside the busy road in front of their small house.

"It's not what most gardeners would imagine," he said.

> *The entire plot is a raised mound of bricks and rubble that arrived after my neighbor built an extension to his house. Thanks to the arrangement of the houses and the road, the prevailing winds blow strongly across the front garden where I grow Tibetan borderland and some western Himalayan species because they appreciate the wind and the extra sun. In essence, it's a wind tunnel with a layer of basic clay and some added grit. But it's a great garden to show just how easy it is to grow the rare and exotic plants that I bring back from afar.*
>
> *Then there's the Bad Garden which faces north and lives with an entirely different climate. Here, with shade and moist, cool conditions, plus extra watering during dry summers, I am now cultivating many* Meconopsis *and unusual primulas.*
>
> *The garden now contains several hundred Himalayan plants and has been designated the United Kingdom Station of the P. N. Kohli Memorial Garden (see page 171).*

His expeditions allow him the right of travel to bring back for gardeners—who do not have the time or inclination to climb mountains or trek across deserts—plants never seen or, at best, little known in garden cultivation. While the eastern Himalaya is monsoon country, Kashmir, the spot that Mr. Chadwell loves the best, has climatic conditions similar to much of England and a large part of the United States. According to Mr. Chadwell, the correct term is *Himalaya*, from the Sanskrit for "abode of snow," so the *s* is not needed.

> *It has severe winters and plenty of sun in the summer—and much of the area is relatively dry. Ladakh, on the western edge of Tibet, has many of the world's highest-flowering plants. The beautiful sandwort,* Arenaria bryophylla, *has been seen blooming at 6,200 meters or 20,000 feet.*
>
> *And you can't always believe what others say about the weather. Dalhousie in the western Himalaya is not far from Dharamsala, the present base of the Dalai Lama, fully exposed to the monsoons. The area registers an average rainfall of about 101 inches (254cm) a year while Leh, the capital of Ladakh 149 miles away, usually sees an average of only 4 inches of rain a year. Other parts of Ladakh are not quite so arid, in particular the Suru and Zanskar valleys. But on my first visit there in 1980, I asked a local when it had last rained and in the style of the White Rabbit from* Alice in Wonderland, *he promptly looked at his watch and*

proclaimed it had not rained for two years. That night it rained, not quite a downpour but wet enough to penetrate the inadequate tent of the expedition's leader.

Politics also enter into Chadwell's expeditions. The list of seeds found in the 1990 issue of *Himalayan Plants Seed* begins with the statement,

Due to serious political unrest supplies of seed from Kashmir have been curtailed this year. Therefore, a modest seed list has been compiled instead of the normal illustrated catalogue. There appears to be no easy solution to the problems in Kashmir and so the situation is likely to be difficult for some years to come.

But when dealing with collecting, Chadwell said,

Restraint must be shown by all visitors, especially those keen on plants. It's not appropriate for the average visitor to dig up plants. The removal of one plant by one collector may not make any difference, but when thousands of collectors take thousands of plants, it isn't long before the rare becomes the extinct. There are times when the removal of live plants can be justified but that decision must be left to trained botanic staffs, to formal expeditions that are staffed with people who have the expertise and facilities needed to allow plants to survive. The camera and the seed packet are a much better way to go.

In addition to his involvement with bringing back seeds for his own business and being one of the founders of the Sino-Hymalayan Plant Association, Chadwell also sells shares in seed expeditions. A share might sell for $50 and allows the investor the choice of a number of rare and unusual seeds, long before enough are cultivated to go on the market.

A typical collection expedition was formed in 1990 to journey to Thak Khola of Nepal. The team consisted of Chadwell; Alastair McKelvie, past editor of the *Journal of the Scottish Rock Garden Society,* who recently retired as deputy principal of the North of Scotland Agricultural College; and through the assistance of an English friend, Diana Penny Sherpani, who runs a specialist trekking company, and as her name suggests, is married to a Nepalese sherpa. We also secured the services of an experienced guide, Kalden.

"I cannot overemphasize," said Mr. Chadwell, "the importance of local guides and collecting assistants. In the past they were given little recognition but often did much of the actual collecting when the famous western plant explorer was not even in the field or the country at the time. Many of the best Himalayan plant introductions of the past, which still appear on rock garden society show benches around the world, were gathered by individuals who received no recognition. This even includes such gems as the blue form of *Paraquilegia anemonoides* (*P. grandiflora*) originally introduced from Bhutan."

The team was in the field between August and October with a mission to gather seeds of horticultural merit from species that grow in a wide range of habitats, including forests, alpine pastures, grasslands and scrubs, stream sides, and high passes and cliff edges. And Mr. Chadwell points out that an expedition to Nepal that takes an interest in botany or horticulture must approach the correct authorities, collect duplicate material for the National Herbarium and the Kathmandu Botanic Garden, and finally, obtain formal authorization to collect and transport seed and herbarium specimens out of the country by the director general of His Majesty's Department of Forestry and Plant Research. Abiding by such rules and regulations is only right and proper. According to Mr. Chadwell:

> *Border sensitivity restricts access all along the Himalaya and some regions can only be entered with special permission. Approval is difficult to obtain at the best of times and it certainly wasn't feasible for this particular venture. But there were so many places sufficiently rich in plants that could be readily reached that we decided to go.*

And if you think the life of a seed collector is easy, read the following entry from Mr. Chadwell's notes:

> *My first day on trek was exhausting. In the western Himalaya you normally begin walking at 2100 to 2700m. Nepal, however, treats you to a 1000m start in very hot [and] humid conditions. My discomfort was not helped by a young, fit and fully acclimatized guide suggesting I must weigh 200kg! On the main paths, stone steps, albeit of varying sizes, are commonplace, aiding rapid ascent [but] care must be taken as they are often very slippery. A trusty walking stick proved invaluable. Whilst a slip is unlikely to be fatal, twisted ankles or damaged limbs are none too helpful with thousands of feet of ascent and descent ahead.*
>
> *A further complication is leeches. The average trekker is hardly troubled by them, being something of a novelty [but] they proved more than a minor irritation to me. In the cooler, normal trekking months they largely disappear and provided that you stay on the main tracks, few reach their targets. Once, however, you submerge into the undergrowth in search of plants, it's a different story. They definitely won during my first few days, resulting in many bloody wounds all over my anatomy. In the leech-free western Himalaya, I am accustomed to making methodical detailed field notes and gathering seeds on my hands and knees. There are penalties involved in Nepal. Being attacked by thirty or more of the little buggers is something of a distraction and you learn to answer a "call of nature" very speedily.*

Among the plants that Mr. Chadwell expected to find seed in sufficient amounts for collection were *Meconopsis paniculata*, a species that forms evergreen rosettes and bears yellow flowers; *Lilium nepalense*; *Chlorophytum nepalense*, the Himalayan member of the Saint Bernard's lily; *Rhododendron barbatum*, a plant that in Nepal becomes a sixty-foot tree; and other treasures like *Saussurea graminifolia*, a strange but beautiful member of the daisy family; and a number of unusual members of the saxifrage family.

For a standard share, approximately thirty-five to forty packets that concentrated on perennials and alpines would be distributed, including a few smaller shrubs and climbers. In essence, a shareholder could really become the first gardener in his or her area with a really unusual or rare plant.

Prospective members are then asked their preferences among perennials, shrubs, climbers, alpines, woodland species, high alpines, plants for a peat garden or a bog garden, trees, tall shrubs, and subtropical species that would require greenhouse cultivation.

But for those gardeners who wished not to buy shares, the local collectors provide a general seed list. These lists always contain mention of seeds from some beautiful plants. Consider the Kashmir form of the Himalayan musk rose (*Rosa brunonii*), a vigorous climbing rose that produces masses of scented white flowers in June—not found in your usual rose catalog—or *Codonopsis clematidea*, the easiest to grow and one of the most beautiful members of this genus of tall rockery plants that produce nodding bells with shiny, skim-milk blue petals, easily missed in a colorful garden. But inside the flower bell, there are lovely markings of purple and orange, markings usually only admired by bees and other insects. I've grown this plant in my old Zone 5 garden, where with very sharp soil, it bloomed every summer.

THE FABLED HIMALAYAN BLUE POPPY

Most seed lists from Mr. Chadwell (and the various rock garden societies) always include the west Himalayan blue poppy (*Meconopsis aculeata*), noted for its spikelike clusters of sky blue and wafer thin petaled flowers on bristly stems, plus the only such poppy I ever had any luck at flowering, *M. horridula*. The lovely petals vary from blue to claret but are so named because the stems are covered with small thorns. With all the elán of a Dr. Johnson, Mr. Chadwell mentions that the plant is "widely grown in Scotland." Said Mr. Chadwell,

> *The genus* Meconopsis *has a magical association for both gardeners and non-gardeners because the electric blues of well-grown specimens are an*

unforgettable sight. The phrase Giant Blue Himalayan Poppy is often used and usually the reference is to M. betonicifolia, *which strictly speaking, hails from Tibet. But there are many other, equally beautiful species that can be grown and most hail from the monsoon climate of the eastern Himalaya.*

There is one species from the western, dried region named Meconopsis aculeata. *The species is distinguished by its blue-gray leaves which are lobed and dissected, and it is sparsely covered in pale straw-colored, prickly hairs. The size of the plant varies from dwarf at three to nine inches high, to well over two feet. The flower color ranges from sky blue to mauve, and there are some white forms.*

According to Mr. Chadwell, it's not a particularly easy species to grow. But he advises that James Cobb in his book, *Meconopsis* (Portland, Ore.: Timber Press, 1989), suggests that it requires a damp spot that does not become waterlogged in the winter (a requisite for most plants from mountain areas). Mr. Cobb suggests a novel growing method that starts with a twelve-inch-deep container with lots of good drainage material at the bottom and a plugged hole. This plug is removed in the winter, and the tub is covered. As with all meconopsis, they resent being kept overly long in pots and like to get their roots into a soil that is laced with a good humus content. Plants are winter dormant and monocarpic (they seed and die after flowering) but usually set plenty of viable seed. And remember that slugs can always be a problem with these plants.

Mr. Chadwell found that the color forms of meconopsis in Kashmir are typically sky blue to deep blue. On his first visit in 1983, he found a large population of these plants growing on cliffs and ledges above Mount Kolahoi's glacier. He explains:

Moving southeast into Himachal Pradesh, you will find a great deal of natural variation in size, with dwarf forms being widespread and colors often containing purple or mauve and pinks that range from delightful deep colorations to some rather dull and muddy ones.

On the Rohtang Pass, the plant grows in crevices on cliff faces, wet ledges, and I even found a dwarf plant about three inches tall that grew on top of a mossy boulder.

ABOUT BERGENIAS

If you buy a bergenia for your garden, chances are it will be a cultivar of a garden hybrid, *Bergenia* x *schmidtii*, the largest, most vigorous, and most common bergenia in cultivation. But Mr. Chadwell describes five in the *Sino-Himalayan Plant Association Newsletter* for February 1993.

Bergenias, often called elephant's ears, are grown by gardeners not only for their bold foliage, but also for the bell-like flowers of red, pink, or white that droop at the ends of long stalks in early spring.

Among the species that he has featured are the winter begonia or *Bergenia ciliata* (*ciliata* refers to the hairs found on the leaf margins and, in some cases, the surface of the leaf); a form called *ligulata*, found growing on shady cliffs in Kashmir at elevations of six thousand to ten thousand feet (eighteen hundred to three thousand meters); *B. purpurascens* with lovely purple flowers; and *B. stracheyi* that bears numerous white flowers that turn to pink with age.

OTHER HEROES

Many of the modern-day plant hunters who have included the Himalaya in their wanderings have become aware of the need to conserve this unique flora in its natural habitat, and their voices can be heard about the need for conservation in India as well as in Western countries. One of the aims of Mr. Chadwell and the Himalayan Plant Association is to establish the P. N. Kohli Memorial Botanic Garden.

The garden would commemorate Prem Nath Kohli, an Indian forestry ranger, horticulturist, and conservationist who introduced many Himalayan plants into Western gardens, both through the Indian firm of P. Kohli & Co. that he established in 1928 and with the Maharajah of Kashmir when requests for seeds or plants of choice alpines came from around the world. English royalty was no exception in the 1920s when Col. Clive Wigram, Privy Seal to King George V, approached the Maharajah for primulas and gentians. *Primula clarkei* and *Gentiana kurroo* are two outstanding rock garden plants that Kohli introduced.

Unfortunately, the conventions of that day prevented the highly talented forest ranger from undertaking serious botanical research (sadly, a situation that often still exists today), but Mr. Kohli wrote for newspapers and specialist bulletins.

Originally, the garden was to be established within the main Kashmir Valley, but events during 1990 changed those plans.

According to Mr. Chadwell:

> *At the beginning of August Mr. Virender Suri, husband of Kohli's daughter, was shot dead by Moslem militants in Srinigar, Kashmir, an event that took place at their home in front of the family. Other members of the family had urged them not to return to Srinigar after a visit to Delhi that summer. But despite the dangers, Mrs. Urvashi Suri, proprietor of P. Kohli & Co., felt a strong obligation to her customers, and disturbances in Kashmir were commonplace. What made the shooting worse were the repeated assurances of neighbors that they would not be*

harmed. Their home was close to the old part of the city, the focal point for most of the violence. Local men carried out the execution simply because Mr. Suri was of the wrong religion.

After fleeing with the body for a funeral in Delhi, it has not been safe to return to Kashmir. As if to seal any question of carrying on in Kashmir, a more distant relative, who had been watering plants at their nursery and keeping an eye on the house, was hung a month later. So at this time the favored area for such a garden is Dehra Dun. Donations will be held in a British Deposit Account until such a time as land is purchased and materials required.

CLUMPING YOUR SEEDS

Contrary to the conventional wisdom concerning seed sowing, Mr. Chadwell has had almost universal success by sowing it thickly, then planting out the resultant mass of seedlings straight back into a convenient hole in the garden. Of his methods, Chadwell notes:

This method has resulted largely from a lack of space because my garden is remarkably small, plus a lack of time, or more likely, a lack of patience. The only real defense I can claim is my desire to go against accepted wisdom based on many years of observing plants in the wild where clumping seems to be the method that nature employs.

In the wild, seed is never dispersed in neat rows, instead most seedlings seem to clump in one spot. Furthermore, seedlings like company and perhaps this might create favorable microclimates. Certainly this is a better environment for growing seedlings than a bleak, open landscape. Of course you eventually need to thin out the clumps in some cases, but I feel that the minor problem of doing this when your plants are growing vigorously, is a problem most gardeners would be happy to face. Also, at this stage in development, root systems are large and strong, so little damage is done when easing them apart, even those which appear to be tangled up.

Mr. Chadwell also knows of another seed-sowing method that he encountered in Toronto, Canada. Barrie Porteous, an exiled Scot, is a serious grower of unusual plants, but, like so many others in his part of the world, he suffers from living in a climate with lengthy winters and many months before the snow melts enough to allow work to begin at his main garden located at a cottage to the north of the city. Although he sows seed under artificial light, for seeds that he suspects might need a chilling period, they are sown in pots, then sealed in plastic bags, and the whole affair is placed in a domestic refrigerator! Even allowing for the greater capacity of North American fridges (English and other out-country refrigerators are much smaller than American models), he rapidly runs out of space.

In a more scientific vein, Chadwell claims that research indicates that some young seedlings benefit from the nutrients produced by the decaying parts of parent plants—certain symbiotic relationships between funguses and trees are now well documented—so perhaps one day a scientist will actually come up with an explanation for his suggestion. After all, the growth of terrestrial orchids was something of a mystery until relatively recently.

He also draws a parallel with landscaping in shopping centers and malls. How often do we see relatively large specimen trees planted out at great expense, only to have an abysmal survival rate (even allowing for the odd acts of vandalism)? It's not only unattractive, but also a poor return on investments. He also asked,

> *Wouldn't it be far better, to plant a mix of sizes and types of plants with ground covers, climbers, small bushes, larger shrubs, and small trees? It would certainly give a more natural look and the seedlings of larger plants would be protected and nursed along in their early days.*

THE PAPER TOWEL GAMBIT

I'd heard about germinating seeds on paper towels but never tried the technique. Then in the *Sino-Himalayan Plant Association Newsletter* (February 1993), I read an article by David White about growing seeds without pots. Said Mr. White,

> *For years, I put off experimenting with germinating seeds in paper towels because I had heard stories about rotting seeds and thought the conventional methods were fine. But then I was overwhelmed by the number of seeds I had to sow and even using two-inch pots the space required was formi-dable. The proportion of sowings that germinated was also beginning to disappoint me, and I did not know if this was due to the compost mix I was using or other factors such as the seed rotting over the winter. When germinations did occur in the pots, it was not always clear if it was of the original seed sown, or of weed seeds blown into the pots in the cold frame. I also have big problems with slugs, snails, and other creepie-crawlies, and I'm sure that many a germination went unobserved as some gastropod enjoyed a midnight feast.*

So last November, much to the amusement of Mrs. White, Mr. White piled a heap of paper kitchen towels and food bags in the supermarket trolley and headed home to experiment with seed collected from previous expeditions.

His first problem was to find a pen that if used on paper towels would not wash out or run when the towels were wet. In England, the W. H. Smith Labelling Pen and in America, the Sanford Sharpie Marker would both write without bleeding and would even take on damp toweling.

As to folding and wetting the towels, after producing a heap of soggy towels, Mr. White developed the following technique:

> *First I take a sheet of 2-ply kitchen towel measuring about 11 x 9½ inches—I use the extra thick variety—folded in half, then in half again to make almost a square, and finally in half again, resulting in a folded towel measuring about 3 x 5 inches. The sowing details are written on the outside of the towel, and the last fold opened up. Dip the end of the towel into a bowl of water until the wick action has drawn the water halfway up the towel. Remove from the water and place it on a clean work surface, then fold the half-folded dry back onto the wet half, and using your hands, press the towel from the middle outwards, until the front half is wet. Squeeze any excess moisture out of the towel.*
>
> *The seed is then placed thinly and evenly on one half of the towel, being careful to remove any chaff or debris. Fold over the other half of the towel, making a seed sandwich. Place the towel inside a reclosable plastic sandwich bag, 6½ x 5⅞ inches, then fold the bag in half and place it in a plastic tub. Leave the tubs in a warm room or a refrigerator, depending on the type of seed.*

The seedlings are picked out into pots that are then sealed inside plastic bags to give them a warm and moist microclimate.

As to knowing when a seed has germinated, hold the flattened packet up to a bright lamp or a sunny window. Those seeds that have germinated are immediately evident.

Mr. White's attempts at germination were an unqualified success. He was staggered at the short time it took for most of the seeds to show signs of germination—as short as three days, in some cases—and was immediately faced with the problems of keeping the seedlings alive until spring. He was also amazed at the ease that certain composite seed germinated—despite its being two to three years old. *Saussurea graminifolia,* collected in 1990, took only three to four days; even some *Celmisia* seed germinated well. This last genus is supposed to produce seed with a short life, but White suspects this may not be the case, and it's more likely that much of the seed is unviable but has some viable seed mixed in.

According to Mr. White,

> *Seed of* Saussurea simpsoniana *provided an interesting education. In the spring of 1992, I sowed seed in compost and nothing happened. When*

I carefully examined the result I immediately saw that I had sown the feather pappus, but no seed. I then tipped out the remaining dregs from the seed packet and sowed this on another paper towel. After four days, two of the tiny specs—that looked exactly like mouse droppings—germinated. I also learnt that seeds of S. simpsoniana *smell strongly of a tom cat, a fact that I doubt if many gardeners know.*

Using paper towels proved to be extremely useful for controlled experiments with seeds. When working with *Arnebia benthamii* (a rare plant belonging to the borage family), White opened a seed packet to confront a jumbled mass of flower heads and dried vegetable debris. He carefully sorted out the seed and by accident, when breaking open a gritlike seed, found a soft green center. He then broke open four more seeds and planted these five in a paper towel. He then put five of the hard seeds in another towel. After just three days, the soft-coat seeds germinated and the hard coats just sat, finally germinating a week later.

THE ADVANTAGES:

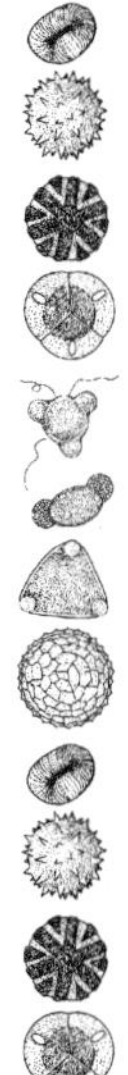
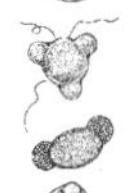
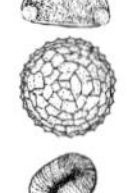

1. The system is a real space saver, freeing valuable cold frames for growing seedlings and plants. Around forty sowings can be kept in that plastic ice-cream tub, and these can be stacked easily. It's an especially useful technique when used with seeds that require many cycles, sometimes years to germinate. Using the standard technique, pots are soon smothered in mosses and liverworts. If the towels get messy, it's a simple job to resow on a clean towel. And the raw materials are easy to recycle.
2. Examining the seeds' progress is clean and easy. No more floors will be covered with grit or sand as you look for signs of germination.

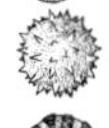
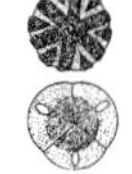

3. Weed seed contamination is almost eliminated.
4. The technique is used to test seed viability.
5. Potting the seedlings is easier with the paper towel technique. Seedlings are easily separated from other root systems without becoming a tangled mess.

"Of course," said White,

many people have had problems when growing seeds with paper towels and keeping seedlings going in mid-winter is not easy regardless of what technique is used. But I've made a number of initial observations: Timing is important, and I have had much greater success by pricking out the seedlings immediately after the leaves have emerged from the seedcoat. But you cannot delay too long or the root systems become oxygen starved. Those that I have pricked out with the seed coat on have tended to be unable to throw it off, and have rotted. However, this deficiency in the seed coat could be due to the large temperature difference the new seedlings experienced when moved from snug and warm towel pots to my unheated conservatory.

The technique is by no means new and has been used for years in botanic gardens, with seed sown on agar in petri dishes. And as a non-gardening neighbor pointed out, it's also used in playgroups to make cress hair or clay animals that sprout green fur. It's not well established with most gardeners but if my initial results are any guide, we have been missing a good trick for many years.

JIM AND JENNY ARCHIBALD

It's difficult to throw away one of the Archibalds' seed lists because each issuance contains news from these intrepid seed collectors. I began my collection in 1986 when Jim and Jenny pulled a caravan (or trailer) to the eastern part of Turkey where it assumed the status of a base camp. They had a good reason for this plan because they wished to look around and collect in a relatively small number of areas, but in more detail, rather than make their tortuous way and collect as they progressed. Said Jenny,

In Greece we left our caravan with Greek friends while we made side trips to Helmos and Parnassos, then it was straight out to the Turkish town of Van about 50 miles due west of the Iran border to leave our home for over two months, deposited in a field beside Lake Van.

We made a great many bulky collections and to be able to get rid of them while traveling was sheer bliss. Moreover, to deal with the work of seed-cleaning, drying, and organizing dried specimens, identifying collections, and all such ancillary work which occupies rather more time than actually gathering the original material in relatively spacious and comfortable surroundings, made life both easier and much more efficient.

In spite of scheduling their activities in great detail, they invariably set off each year with the intention of collecting many species, whose homes they never find enough time to visit. Their only consolation is that these plants remain to tempt them back on some future date and will eventually provide everybody on the list with new and exciting material, the lifeblood of their work. Within the planned scope of this year's journey were such incredible and desirable species as *Potentilla oweriniana*, *Rhodothamnus sessilifolius*, and *Campanula bakkiarica*. But they aren't listed because they failed to find them; in fact, they didn't attempt the search.

> *We are essentially under the pressure of achieving maximum productivity: we must always select localities that are both the easiest access and will yield the greatest number of worthwhile collections. A species like* P. oweriniana *means devoting at least three days to the attempted collection of what might be only one item and even then the attempt might end in failure.*

In 1987, the Archibalds visited the American West, collecting seeds in the Rocky Mountains, stopping at Elk's Park on Pike's Peak and all around Santa Fe, and found that everyone in Wyoming seems to drive a pickup truck with a gun rack along the back of the cab.

> *We found no great nomadic tribes and the Plains Indians can no longer follow the herds of bison. The "snow-birds," a gentle race of elderly people, follow the sun, often waiting in the mountain areas for their next Social Security check so that they can fill up the tanks of their thirsty R.V.'s and move to the South, or perhaps they will spend the money on sugar to keep their humming-bird feeders full. Many of the travelers we met were interested in the local flowers and what we were doing. Some just enjoyed them: One pressed flowers and one actively collected, filling the shower room of their large motor-home with an assortment of artemisias, cactuses, and even a large buffalo berry (*Shepherdia *spp.). And at times, we felt more remote than one can ever feel in Europe or even Turkey—America is a big country.*

In 1988, my Archibald seed list reported the establishment of the Archibald Base Camp in rural Wales. Their address is now Bryn Collen, Ffostrasol, Llandysul, Dyfed, Wales.

Their new home lies to the far west of Britain, on a north-facing slope near the base of a typical little Welsh valley that plunges steeply down from the uplands at an altitude of 220 meters (746 feet) and that looks down at a little stream called Nant Collen.

OFF TO ECUADOR

I lost track of the Archibalds when we moved from New York to North Carolina, and my seed list collection picks up again with the listing of seeds collected in Ecuador, in July 1993. The following is a quote from that newsletter:

> *The climate cannot be summed up briefly. It is incredibly diverse and every mountainous area—if not every peak and valley—has its own*

micro-climate. Warm, wet air rises from the Amazon Basin in the east and the Pacific in the west to produce an overall wet mountain climate though it can be very dry locally in interandine valleys lying in rain-shadows. Summer and winter are replaced by irregular dry and wet seasons. Temperatures fluctuate more between day and night in the dry season because of a lack of clouds but it can be much colder overall during a wet season because there is no daily sunshine. For most mountain plants in Ecuador, the wet season is winter and flowering peaks at the beginning of the dry season in many cases. This seems alien to gardeners used to conventional climates but the overall situation at higher altitudes (and it is only these which concern us here) is not so very different to cool, temperate climates, such as the British Isles, parts of New Zealand, Southeast Australia, and Northwestern North America, where winters are not really very cold or summers very hot. All the plants in such climates from which we have collected seed will be very much happier outside during the summer. The treatment of equatorial Andean species in winter will depend on the altitude at which they grow and, rather than quote statistics, we feel it may mean more to gardeners if we look at three broad altitudinal categories and the foreign plants which are at home in them. It is only by adventurous and experimental interpretation of these that we can proceed.

2500 to 3500 meters—Quito, at 2850 meters has a climate which has been described as eternal spring, though in the wet season it can be quite a chilly spring that is rather like a Mediterranean winter. In the gardens Cyclamen persicum *grows and flowers non-stop, argyranthemum cultivars thrive and Watsonias naturalize with ease. Between 2500 and 3000 meters, plants will be unlikely to experience serious frosts and they should be grown frost-free in winter. Most of the population—Ecuador is the most densely populated country in South America—live in the intermontane basins around this altitude and the natural vegetaion is almost entirely removed. Approaching 3500 meters it is noticeably cooler.*

3200 to 5000 meters—This narrow band, which varies somewhat according to the locality, was, to us, the most interesting zone. About here the montane scrub grades into grass paramo, sometimes abruptly, sometimes through dwarf elfin forest. We should guess plants from here would be most successful in cool, moist climates, such as on the west coast of Britian, provided frosts were not severe—Cornish and Irish material. Here in some areas Digitalis purpurea *is naturalized. These are plants for experimentation; some might be hardier than we imagine. Cool summers will suit them.*

3500 to 5000 meters—At the lower levels there is paramo, the high altitude moorland so characteristic of the North Andes. The dominant plants are grasses and the other species occupy specialised niches, such as wet sites or stony sites. Little in the way of new species occurs above

> *4300 meters.* Nototriche *starts at about 4000 meters. Only the odd* Draba *extends to over 5000 meters; there would be too little time uncovered by snow for most plants above 4500 meters. The snow-capped summit of Chimborazo is 6310 meters, Cotopaxi is 5897 meters, and Cayambe is 5790 meters. The problem in growing alpines is much more likely to be warm summers than cold winters. We should advise growing such plants outside throughout the year in the United Kingdom; many should thrive on British winters, autumns, and springs. This is not the Peruvian or Bolivian altiplano; Ecuadorean alpines survive rain, hail, sleet, and snow and skies which are often overcast with cloud. Long, hot summer days are another matter but the alpine specialist has always been optimistic and inventive. Take heart and bear in mind what we have written here is an appalling generalization—homely northern conifers have been planted in the paramo, which comes down to 3200 meters to 3300 meters, west of Cotopaxi. It looks just like the Scottish Highlands.*

Just to give the reader an idea of how the Archibald seed list is documented, here are ten entries from the Ecuador seed offerings. If they don't make an adventuresome gardener's mouth water, what will?

> Bomarea *spp. This magnificent genus has remained little-known in cultivation and one of our aims was to make seed collections from some of the higher altitude species. Belonging to the* Alstroemeriaceae, *these are mainly climbing tuberous-rooted perennials with regular flowers (unlike* Alstroemeria*) centered on the North Andes but extending north into Central America and just entering northern Chile in the south. They are particularly diverse in Columbia and Ecuador and have a great altitudinal range—from lowland jungle to almost 4500 meters. There are about fifty species [*Hortus Third *lists 100] and they are being revised at present for the* Flora of Ecuador, *so we have desisted from guessing at names. In several cases, we had the impression that populations were intergrades or hybrid-swarms, such was their variability. They rely mainly on hummingbirds for pollination, so many have orange-scarlet flowers, and often to help in seed dispersal, capsules open to display seeds with a fleshy, viscous, bright-orange skin.* Bomarea *spp. Equador, Napo, Papallacta. 3100 meters. Margins of montane forest. July 1, 1993, climbing to 2 to 3 meters with heads of up to fifty flowers, unspotted with scartlet outer segments and orange inner ones.* Bomarea *spp. Ecuador, Imbabura, northwest of Laguna Mojanda (south of Otavalo). 3700 meters. Dense elfin forest at edge of paramo. July 9, 1993, the highest collection—and largest species—stout, downy stems climbing to 2 to 3 meters with leathery leaves and enormous heads—we counted 120 flowers in one of huge, chocolate-spotted, apricot-yellow bells. The whole genus is magnificent but this is the* creme de la creme.

Calceolaria hyssopifolia. *Ecuador, Pichincha, southeast of Cayambe. 3500 meters. Open areas, along eroded, stony banks. July 3, 1993. About 50 centimeters high (20 inches), shrubby endemic to North Equador. Narrow, sticky, leathery, bright green leaves and cymes of big, rounded bubbles with smugly closed mouths—white shading to pale yellow. A lovely thing.*

Chuquiraga jussieui (C. insignis). *Ecuador, Cotopaxi, northern slope of Volcan Cotopaxi. 4300 meters. Paramo on open north- and west-facing slopes. July 12, 1993. The amazing shrubby composite, characteristic of the highest altitudes in Ecuador extends from southwest Colombia just into northern Peru and then reappears in southern Peru and adjacent Bolivia. Here at about its highest elevation, it's 50 to 80 centimeters high with erect woody stems clothed in little stiff, spine-tipped, imbricate leaves and producing large, terminal everlasting heads of shiny, papery, bronze-orange phyllaries from which the yellow-orange flowers appear with exserted scarlet styles. Both bizarre and very beautiful.*

Cleome *spp. Ecuador, Canar northwest of Canar to Chunchi. 2400 meters. Dense montane scrub on steep, wet slopes. July 23, 1993. An erect shrub, 2 to 3 meters high, with the characterisitic, strange yellow flowers followed by bizarre fruits like hanging sausages covered in pink felt. From this altitude for frost-free conditions only.*

Ephedra americana. *Ecuador, Cotopaxi, northern slope of Volcan Cotopaxi. 4200 meters. Exposed stony areas. July 12, 1993. Tight, prostrate mats of greyish, interlacing stems with a profusion of orange-scarlet fruits.*

Eryngium humile. *Ecuador, Azuay, Paramo de Tinajillas, west northwest of Nabon. 3200 meters. Open paramo among grass tussocks. July 22, 1993. A sweetie tiny rosettes of entire, spine-edged leaves and branching stems of heads with silvery-white bracts surrounding a dome of navy-blue flowers. Like a miniature, papery blue and white Astrantia maxima. Very polymorphic (this is a good form), it is 20 to 30 centimeters high. Try it in scree or raised bed.*

Gaultheria. *Ecuador is rich in Ericaceae with many genera unfamiliar at northern latitudes—we were too early for seed of the magnificent* Befaria *spp.—the rhododendrons of the Andes. We have a few other ericaceous collections which we shall include in our next list when more complete determinations on our dried specimens have been received. For the moment we confine ourselves to those which appear to be* Gaultheria *spp., a genus which now includes* Pernettya, *some may be Vaccinium as we are none too skilled at distinguishing these in fruit. All should grow in the alpine house or cold greenhouse and several may prove hardy ouside in the United Kingdom and the warmer parts of the United States.*

Gaultheria *spp. Ecuador, Imbabura, northwest of Laguna Mojandra (south of Otavalo). 3700 meters. Edge of dense montane scrub on tree line. July 28, 1993. Shrub of about 60 centimeters with masses of large pearl-pink to white fruits.*

Gentianella *spp. Ecuador, Cotopaxi, nothern slope of Volcan Cotopaxi. 4200 meters. Exposed, stony areas in volcanic ash and debris. July 12, 1993. Hummocks of tight rosettes of tiny leathery leaves covered with flowers like wide open crocuses in pink to carmine with occasional whites—breathtaking. Try it in scree.*

Ourisia chamaedrifolia. *Ecuador, Pichincha, southwest of Volcan Cayambe. 4200 meters. Fissures on moist cliffs. July 3, 1993. Prostrate, rooting stems with small, rounded, crenate leaves. Tubular flowers, held horizontally on 1 to 2 centimeter pedicels, in a lovely old velvet red with buff lines in the throats. An exquisite, little plant.*

Puya *spp. Ecuador, Imbabura, above Laguna Cuicocha (west of Cotacachi). 3200 meters. Among grass and low montane scrub on steep sides of crater. July 7, 1993. Big, spine-edged, evergreen rosettes send up spires to 2 to 3 meters with buds wrapped in brown wool opening to tubular, waxy flowers in opalescent duck-egg blue-green. Amazing thing.*

THE WINTER OF 1994

When they wrote of their plans during the winter of 1994, the Archibalds had no intentions of leaving Wales during the spring and summer after returning from Argentina in March. But they were given the opportunity of going on a seed-collecting trip to southern Turkey. Jenny stayed at home to garden, and Jim went with his friend Norman Stevens to collect.

Jim had not visited Turkey since 1988 and hoped to collect species they had missed in the past. As it turned out, he could not have chosen a worse year because the season had been disastrous for the Southern Anatolian steppe flora and not much better in the mountains of the Southwest. Crocus seeds, for example, had not set or had dehisced too early. Of course, a number of plants were found, but their condition was described as poor, and the amounts of collected seed were miserably small—usually far too few for offering in their list.

What was disconcerting for Jim was that so many changes had occurred in Turkey since his last visit. They frequently failed to find some of his favorite localities, and when they did, they often found everything changed. Jim's favorite site for *Fritillaria alfredae* was completely overgrown by oak scrub, and a hillside near Gaziantep that was rich in a bluish form of *Iris sari* had been planted with pistachios. They completely failed to find the area where Jim and Jenny had collected *Muscari macbeathianum,* after spending a morning looking for it. Said Jim,

> *As this is the only known site and we are the only people who know where it is, this is somewhat embarrassing. But there are many new roads and many more surfaced roads, with the old roads being abandoned and falling down the mountain sides. It's probable that this was the last summer that it would be possible to drive over the splendid Irmasan Pass north of Akseki.*
>
> *But on the other hand, new roads open up new localities and there continue to be vast areas of Turkey that are very difficult to reach and are still unknown botanically. The people, of course, were as delightful as ever. The booming then fragile economy accompanied by hyper-inflation, and the increasing military influence on their lives have done little to suppress Turkish ebullience.*

And in 1996, we will see the Archibalds' smiling faces once again on the lookout for seeds that will ultimately find their way to the gardeners of the world, who can enjoy something new instead of the same old thing in the rock garden or perennial border.

CHAPTER 15

Soil Mixes, Equipment, and Planting Out

In 1969, when my wife and I moved to an old country house in Cochecton, New York, we found many things of interest stuffed in the walls. These stuffers were not meant to hide family histories but to close off holes where winter winds would blow into the living quarters. Local lore told us that in the 1930s, one of the owners was interested in plants and had a small makeshift greenhouse (described as a lean-to) made of old storm windows that rested up against the back end of the garage. One of the more interesting papers was a carbon copy, labeled *Seeds*, and written in the left margin were the following words, taken from Karel Čapek's *The Gardener's Year*, sometime in 1936.

> *Some people say that charcoal should be added, and others deny it; some recommend a dash of yellow sand, because it is supposed to contain iron, while others warn you against it for the very fact that it does contain iron. Others, again, recommend clean river sand, others peat alone, and still others sawdust. In short, the preparation of the soil for seeds is a great mystery and a magic ritual. To it should be added marble dust (but where to get it?), three-year-old cow dung (here it is not clear whether it should be the dung of a three-year-old cow or a three-year-old heap), a handful from a fresh molehill, clay pounded to dust from old pigskin boots, sand from the Elbe (but not from the Vltabva), three-year-old hotbed soil, and perhaps besides the humus from the golden fern and a handful from the grave of a hanged virgin—all that should be well mixed (gardening books do not say whether at the new moon, or full, or on midsummer night); and when you put this mysterious soil into flower-pots (soaked in water, which for three years have been standing in the sun, and on whose bottoms you put pieces*

of boiled crockery, and a piece of charcoal, against the use of which other authorities, of course, express their opinions)—when you have done all that, and so obeyed hundreds of prescriptions, principally contradicting each other, you may begin the real business of sowing seeds.

I find the previous quote very amusing, especially when checking the advice given by various garden pundits in how-to magazines, newspaper columns, and some books. The truth is that six components are necessary to healthy germination:

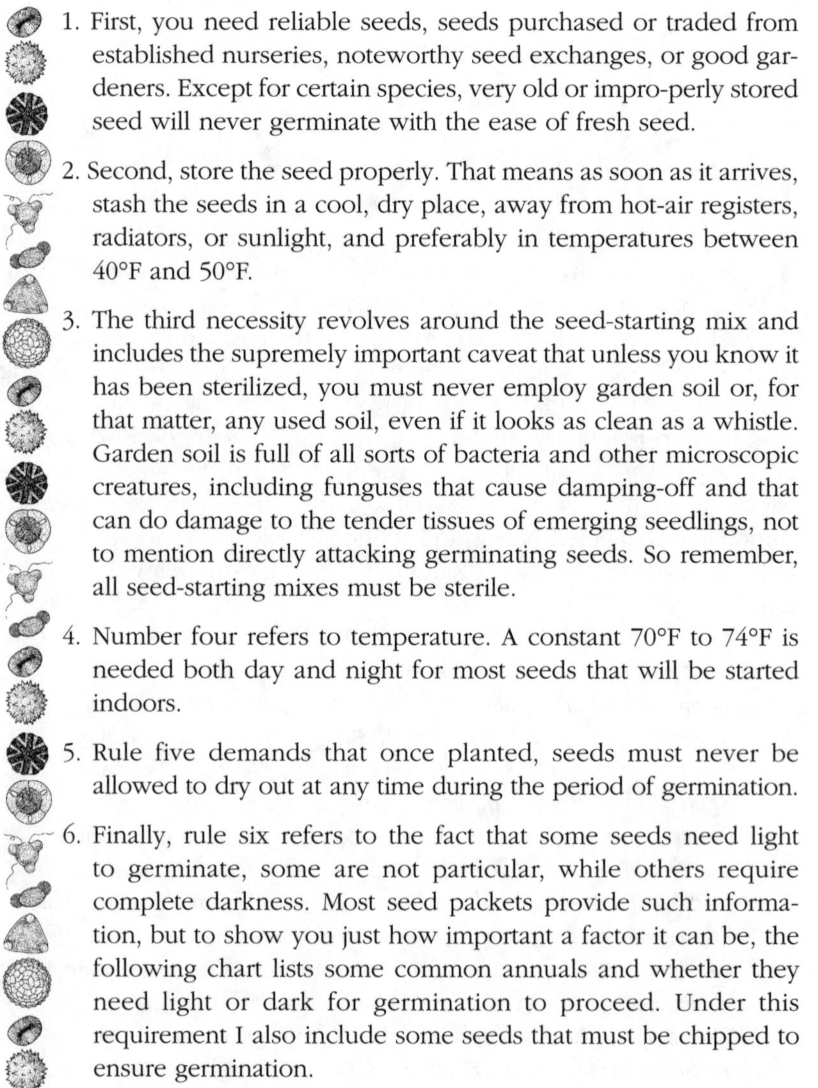

1. First, you need reliable seeds, seeds purchased or traded from established nurseries, noteworthy seed exchanges, or good gardeners. Except for certain species, very old or impro-perly stored seed will never germinate with the ease of fresh seed.
2. Second, store the seed properly. That means as soon as it arrives, stash the seeds in a cool, dry place, away from hot-air registers, radiators, or sunlight, and preferably in temperatures between 40°F and 50°F.
3. The third necessity revolves around the seed-starting mix and includes the supremely important caveat that unless you know it has been sterilized, you must never employ garden soil or, for that matter, any used soil, even if it looks as clean as a whistle. Garden soil is full of all sorts of bacteria and other microscopic creatures, including funguses that cause damping-off and that can do damage to the tender tissues of emerging seedlings, not to mention directly attacking germinating seeds. So remember, all seed-starting mixes must be sterile.
4. Number four refers to temperature. A constant 70°F to 74°F is needed both day and night for most seeds that will be started indoors.
5. Rule five demands that once planted, seeds must never be allowed to dry out at any time during the period of germination.
6. Finally, rule six refers to the fact that some seeds need light to germinate, some are not particular, while others require complete darkness. Most seed packets provide such information, but to show you just how important a factor it can be, the following chart lists some common annuals and whether they need light or dark for germination to proceed. Under this requirement I also include some seeds that must be chipped to ensure germination.

Light or Darkness Needed for Germination of Selected Annuals

Plant	Need	Plant	Need
Ageratum	Light	Hollyhock (annual)	Either
Alyssum	Either	Impatiens	Light
Aster (annual)	Either	Kalanchoe	Light
Begonia (fibrous-rooted)	Light	Kochia	Light
Bells of Ireland	Light	Lobelia	Either
Browallia	Light	Marigold	Either
Calceolaria	Light	Nasturtium	Dark
Calendula	Dark	Nicotiana	Light
Carnation (annual)	Either	Nierembergia	Either
Celosia (cockscomb)	Either	Pansy	Dark
Centaurea (cornflower)	Dark	Pentas	Light
Coleus	Light	Petunia	Light
Coreopsis	Light	Phlox (annual)	Dark
Cosmos	Either	Portulaca	Light
Cuphea	Light	Rudbeckia	Either
Dahlia	Either	Sanvitalia	Light
Dianthus (annual pinks)	Either	Snapdragon	Light
Dusty miller (Cineraria)	Light	Stock	Light
Emilia	Dark	Tithonia	Dark
Exacum	Light	Verbena	Dark
Gaillardia (annual)	Either	Vinca (periwinkle)	Dark
Geranium	Either	Zinnia	Either
Heliotrope	Either		

ORDERING SEED

When most gardeners see a new seed catalog, they metaphorically salivate. The desire to have everything available is diffucult to control. But try not to be smitten with the great pictures in the catalogs. Limit your buying to what you have room and equipment for, remembering that your time is valuable and that the sight of endless trays, all with emerging seedlings, can be an awesome responsiblity.

STORING SEED

When seeds arrive, you'll notice that most commercial packets are made of a heavy, glossy paper or that a few consist of various foils, with instructions for sowing on the label. Seeds from the various societies usually are packed in glassine envelopes, the kind that stamp dealers use to show off their wares. Only the foil packets will keep out air and moisture. So for all-around protection, the best way to keep the seeds is to store them in a closed jar or container. Then keep the container in a cool spot, away from sunlight and heat.

See whether any seeds need stratification. If indicated, now is the time to start that procedure. Open the package and put the seeds in a slightly moist mixture of sterile sand and sphagnum moss. Put the mix in a baggie, with the top tightly sealed and label the date and name. Store in the refrigerator for at least two months. When the correct time has passed, the seeds and medium can be sown as a unit.

KEEPING RECORDS

Always keep track of the name of the seed, the source, the time of arrival, and the date of planting. It's a human failing to think we can remember everything.

I keep a small record book with room for the date, the common and scientific name, the source, the type of compost used, date of germination, and, if needed, a few comments. Those asides can include remarks on erratic germination and whether the seedlings succumbed to damping-off.

If you have a desktop computer (and can train yourself to remember to use it), you have a great tool for record keeping at hand. But index cards and a pencil will also do the job. Either way, if you get seriously involved with raising plants from seed, it's necessary to keep records.

LABELS AND LABELING

Because of my general sorrow, my wife never accompanies me into the garden when spring rolls around. I'm usually sad because no matter where I live or garden, I will always try growing something new, a plant someone sent me from northern Georgia or seeds collected in a part of Chile where the weather isn't as bad as in western North Carolina. I'll give them the old college try, but sometime during the winter they succumb. And every plant or seedling is marked with a white label, labels that become tombstones (or random teeth), all marking experiments that never succeeded.

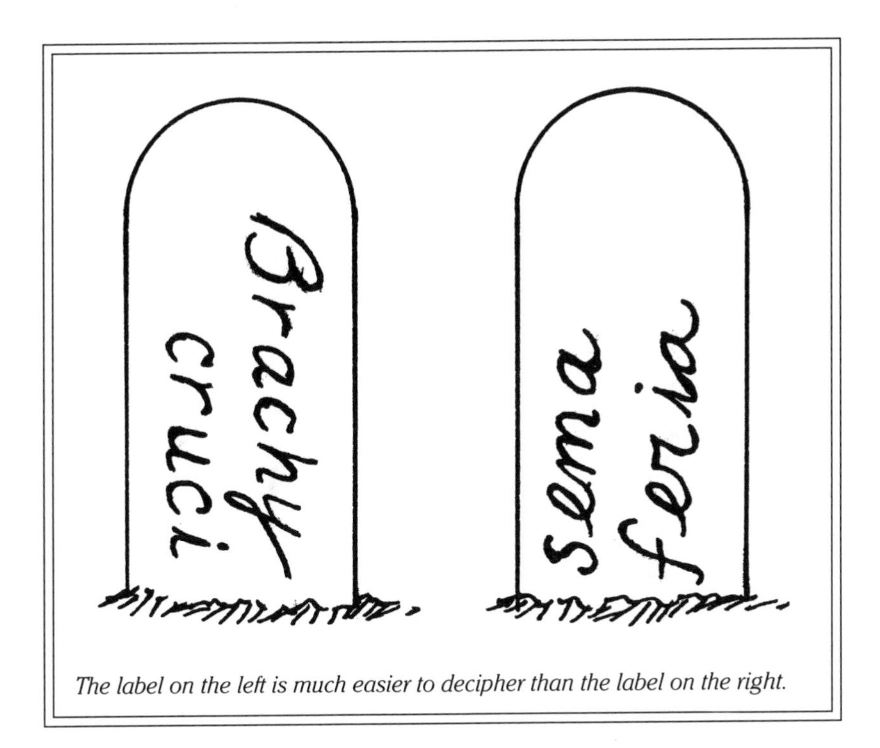

The label on the left is much easier to decipher than the label on the right.

That doesn't mean that labels are unimportant. They are. There isn't a gardener in the history of horticulture who has not said, "Now that's the spot I planted the seeds of ____________________, that's easy to remember," and two months later completely forgets everything connected with that particular activity. I do it continually and have been doing it since I was a teenager.

There are three basic types of labels: wood, plastic, and metal. Wood is getting harder to come by, although you can still use tongue depressors. For me, metal labels are too permanent for marking seeds that might never germinate. That leaves plastic. Plastic labels take pencil or marking pens. I use whatever I can get my hands on. But regardless of which you use, remember one thing: Start writing at the top of the label and write toward the pointed end because it's easier to guess a word from the first few letters than from the last few. *Brachy* . . . *cruci*. . . is easier to read than . . . *sema* . . . *feria*.

And when you have a number of pots in a row, always give each pot a label rather than one label for the head pot. Invariably, pots get mixed up, and you might never know what you have.

DAMPING-OFF

I don't know the derivation of the term *damping-off*, but it's been used for centuries. It's a term that gardeners and nursery men and women

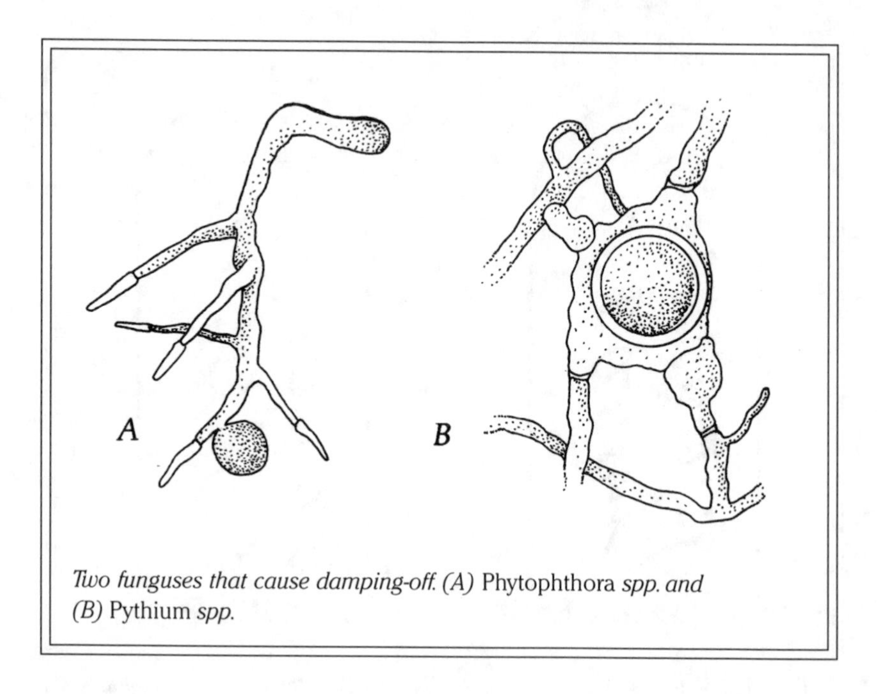

Two funguses that cause damping-off. (A) Phytophthora *spp. and (B)* Pythium *spp.*

use to describe the quick death of emergent seedlings, usually just above the surface of the soil. The conditions that lead to damping-off are an excess of moisture in both the soil and the air, poor light, higher temperatures than necessary for normal plant development, and poor air circulation. Once present, the disease usually spreads rapidly, destroying all the seedlings within a few days.

The funguses that cause the problems are species of *Phytophthora* and *Pythium*, primitive microscopic organisms that belong to a group called the water molds. They love a warm, wet environment and spread from seedling to seedling in moisture. When watched through a microscope, these marauders produce self-propelled spores that move through water until they come to rest on a root or stem. Once there, they attack, just like alien creatures from the movie *The Invasion of the Body Snatchers*, and a thin thread emerges to penetrate the victim's cells. Soon the individual seedlings fall over as though cut down by unseen forest rangers. When temperatures are cool, these organisms move slowly, but at 68°F, they speed about, and damping-off proceeds at a furious pace.

There is another perpetrator of doom, called *Rhizoctonia* spp., that produces a disease called wire-stem and that spreads through the seed mix using fungal threads to choke the life out of seedlings.

These funguses grow inside the seedling's stem and block off the normal cell processes, leading to eventual collapse. Once it starts, it's like the old domino theory, and there is little the gardener can do. Various fungus drenches are used commercially, but they are also dangerous to people and hard to come by.

This spring I hurried along when some seeds arrived from England, and not heeding my own advice, I reached for some old plastic trays and used some cactus soil mix. The seeds were *Aubrieta gracilis.* I put the containers under a plastic lid, set them on the heating cable, and within a few weeks they germinated. But then the scourge struck. If I had checked my references, I would have found that aubrietas, along with salvias, stock, and wallflowers, are prone to *Rhizoctonia* attack. Snapdragons, China asters, cinerarieas, cucumbers, lobelias, and marigolds are also susceptible to *Pythium* and *Phytophthora.*

So what's the gardener to do? Always, always, use sterile growing mixes. Either buy name-brand mixes that are clearly marked sterile or sterilize your own mixes.

And when it comes to other activities, everything should be clean. If you decide to use old pots and seed trays, remember they must be washed in detergent or soap, then carefully air dried. I always add just a dash of a proprietary disinfectant to the wash water. Then, too, use clean water. Never let water sit in a sprinkling can for weeks on end, then turn around and water seedlings. Finally, make sure your seed mix is thorougly damp before you fill the flats, pots, or other containers.

SEED-STARTING MIXES

As mentioned before, pure garden soil is full of microscopic living things, some benign and many more deadly not only to the seedlings, but also to the seeds themselves. Because of these impurities, sterile soil or imitation soil mixes should always be used for germination.

The following materials can be used alone or in combination with one another for seed germination.

Aquarium gravel and bird gravel, found in pet supply stores and supermarkets, is useful in providing proper drainage to a mix, especially for alpines and various desert plants. It's a good substitute if sand, perlite, or vermiculite are unavailable. These gravels should not be used alone because they drain too quickly and provide no nutrition.

Humus is the residue of the compost heap. It is decomposed vegetation that is black, sweet-smelling, loose in texture, but must be sterilized to prevent damping-off.

Loam is the catchall word for good garden soil. It can be used for seeds, but once again, it must be sterile.

Perlite is made from volcanic glass that occurs naturally in globules. It's used like sand but is too light to be used alone and provides no nutrition.

Peat, sold by the bag or bale, is the organic residue of many different plants, including sphagnums that have started to decompose in water. On the acid side, it's very difficult to wet, packs down too tightly, and has little nutritive value. Use only when mixed with other ingredients.

Peat moss is partially decayed sphagnum moss, usually acid, has great water-holding abilities, and like peat, when completely dry, takes an extraordinary amount of time to soak up water. Used by itself, peat moss can actually repel water, so make sure you mix it with other materials before using.

Sand is one of the main constituents of a good mix, especially for cactus seedlings. Use builder's or sharp sand, and make sure it's clean. Sharp simply means that the sand grains are rough to the touch, as opposed to soft sand, which is too fine to be useful. Because it's soft and contains salt, ocean beach sand should never be used.

Sphagnum mosses are found in bogs and swamps. They grow very slowly, the lower parts gradually packing together to from a compact mass that is sold as peat. Sphagnum species have long, hollow cells with a fantastic ability to absorb water. They can usually hold from ten to twenty-five times their dry weight in water for long periods. Sphagnum mosses are not endangered, but many conservationists are against the continual digging up of the peat and the selling of peat moss to the garden market.

The acidifying ability of sphagnums contributes to acid peat formation, a process that greatly inhibits decomposition. Under proper conditions, this chemical reaction leads to such deep beds of only partially decomposed peat moss that wooly mammoths and even humans have been almost perfectly preserved after thousands of years of burial.

Thanks to a cotton shortage during World War I and because of its high absorbency, sphagnum was used to dress wounds. Many sources talked about the moss's having a natural sterility caused by a slight antiseptic quality. Comtemporary research showed that the large sphagnum cells contain a bacteria that is known to either slow or stop certain plant diseases. And this bacteria will spore and survive periods of drying, reviving when the moss is moistened again.

For these reasons, sphagnum moss makes an excellent medium for seed germination and aids in fighting damping-off. For years I would start seeds in sphagnum moss alone, but this proved to be a liability because the moss is sterile and has no nutrition to speak of. All such seedlings needed applications of fertilizer as soon as their true leaves appeared. To save time and effort, I now plant my seeds in a soil mix (see below), but sprinkle shredded sphagnum on top of the mix to a depth of less than a half inch, and I rarely lose anything to fungal disease. Recently, I was careless and lost a whole tray of seedlings, so everybody needs to be reminded about this scourge.

Like all peats, when dry, sphagnums are initially difficult to get wet, and if not kept continually moist, they form a tough surface crust that repels water. Sphagnum is sold both milled and unmilled. Either works well for germinating seeds, but the milled form is easier to handle.

Store-bought mixes are available under many trade names, are generally very good, but are much more expensive than preparing your own.

Vermiculite is a lightweight material made from mica heated to 2,000°F, whereupon it breaks into very small particles. It's often used to grow seedlings, does not pack down, but has a tendency to hold a little more water than seeds demand. It has no nutritive value.

Any of the materials listed above could be used to germinate seeds; after all, seeds fall in the most unlikely places and manage to survive. But nature can afford to waste millions of seeds in order that only a small percentage can germinate—gardeners cannot. Of the various combinations that I've tried through the years, I've found the best seed germination medium to be one-quarter milled spagnum moss, one-quarter vermiculite, one-quarter perlite, and one-quarter sharp sand. If you cannot find milled sphagnum moss, substitute milled peat moss.

A HOMEMADE SOIL STERILIZER

If you truly believe in always making do with what's at hand, there is a way to use garden soil for the germination of seedlings. First, get yourself a large can (once called lard cans) that is about twelve inches in diameter and a foot or so high, and make sure it has a cover. You will also need a piece of heavy galvanized sheet metal, the type that is used to make hot-air ducts for a home furnace.

Set the can on the sheet metal, and using a soft pencil, mark a circle around the can's bottom. Then using metal shears, cut around the circle about one-eighth of an inch inside the mark. Then take a nail and punch several rings of holes in the disc.

Place two bricks on edge in the bottom of the can, and add enough water to cover the bricks. Put the perforated disc on top of the bricks. Fill up the can with soil to be sterilized, and put on the cover.

Temperatures Needed to Kill Soil-Borne Pests

Pest	***30 minutes at temperature***
Nematodes	120°F
Damping-off funguses	130°F
Pathogenic bacteria and funguses	150°F
Most insects and plant viruses	160°F
Most weed seeds	175°F
A few resistant weed seeds and resistant viruses	212°F

When I lived in the country and a trip to a garden center for soil mixes was a big deal, I always sterilized my own soil and began by mixing sharp or builder's sand, half and half with good garden dirt. Because there is an odor involved with this process (baking soil will never win olfactory prizes), my wife allowed the process to occur only in good weather, when windows could be opened to the country air. On damp days, I used a cheap electric single-element heater and cooked the soil in the garage.

When the water in the bottom comes to a boil, steam rises through the punched disc, works its way up through the soil, and eventually comes out from under the cover, which should not be on too tight. Once the steam begins to escape, leave the cooker on for a half hour. When everything has cooled, the sand and soil is ready to be mixed half and half with either granulated peat moss or sphagnum moss.

TEMPERATURE

The single most important reason that many gardeners have trouble getting seeds to germinate is simply that the seed bed is too cool for the temperature requirements of the seed. While many crop plants refuse to germinate unless subjected to temperatures below 65°F, most annuals, perennials, and many houseplants will not begin germination unless they are given temperatures above that number. Lack of heat also works in combination with too much moisture, rotting the seeds before they can sprout. The majority of seeds prefer temperatures between 65°F; and 80°F; the magic number is about 72°F.

There are reasons for these differences. Some seedlings resent the temperatures of spring because they continue to rise as summer approaches. By limiting germination to lower temperatures, they keep germination from occurring until the following autumn, when the environment is more to their liking. Many tropical plants must have the security of germinating to high temperatures, or they might sentence themselves to an early demise. Finally, many perennials begin to go into a state of dormancy below 40°F, so their seeds must be prodded into germination when temperatures are warming rather than cooling.

The typical house or apartment is a place of drafts and fluctuating temperature pockets. The area near a heater or a hot-air duct will be much warmer than around the windows. Near the windows, winter or spring chills can radiate from the glass and easily cause nearby temperatures to fall. Then the moisture evaporating from seed flats can cool the top layers of the germination mix by as much as ten degrees.

And don't forget the water temperatures. Using tap water can quickly give seed flats a chill. When 45°F water was showered on 65°F soil, temperatures dropped almost to that of the water, and it then took three or four hours for the soil temperature to rise again. Seeds are alive and the chemical processes within their seed coats can be adversely affected by such changes.

Far and away the best solution for warming a germination mix is a waterproof, insulated heating cable. Unless purchased from commercial outlets that deal with nurseries, most heating cables sold today come with a built-in thermostat set for about 74°F. All are waterproof. The three-foot cord has a plug, while the thermostat and heating wires are hermetically sealed in vinyl plastic. The cables come in four lengths: 42 watts at 12 feet, 84 watts at 24 feet, 126 watts at 36 feet, and 168 watts at 48 feet; the most expensive is under twenty dollars. The wattage tells you how much they cost to operate. The 12-foot cable uses the power of about a 40-watt light bulb.

To arrange the cable on a utility table or other foundation, begin by putting a layer of Masonite or other insulating material on the bottom. You can even use several layers of cardboard. Place the outside rows of the cable two inches from the edge. Space the adjacent rows of the cable three inches apart. Remember the cables cannot be cut or spliced; they should be spaced evenly by looping back and forth, and they should never cross over each other. Hold the cable in place with electrician's tape.

But if you are unable to get together all the materials to make a propagation box, you will find a sixty-watt electric light inside a box will generate enough heat for many of the more common seeds.

A PROPAGATING FRAME

The following drawing shows a simply constructed, heated seedbed made from easily obtained materials.

(1) 3/4" x 4" planking; (2) 1/2" or 3/4" waterproof plywood; (3) aluminum nails; (4) heating cable; (5) thermostat; (6) heavy plastic sheeting; (7) staples; (8) sand or gravel filler.

The cable is held in place by staples—be careful not to pierce the plastic insulation. As mentioned before, place the outside rows of the cable two inches from the edge of the box. Leave about three inches between the adjacent rows. Never let the cable touch the thermostat or cross another cable because it not only will prevent even distribution of heat, but also may shut the thermostat off.

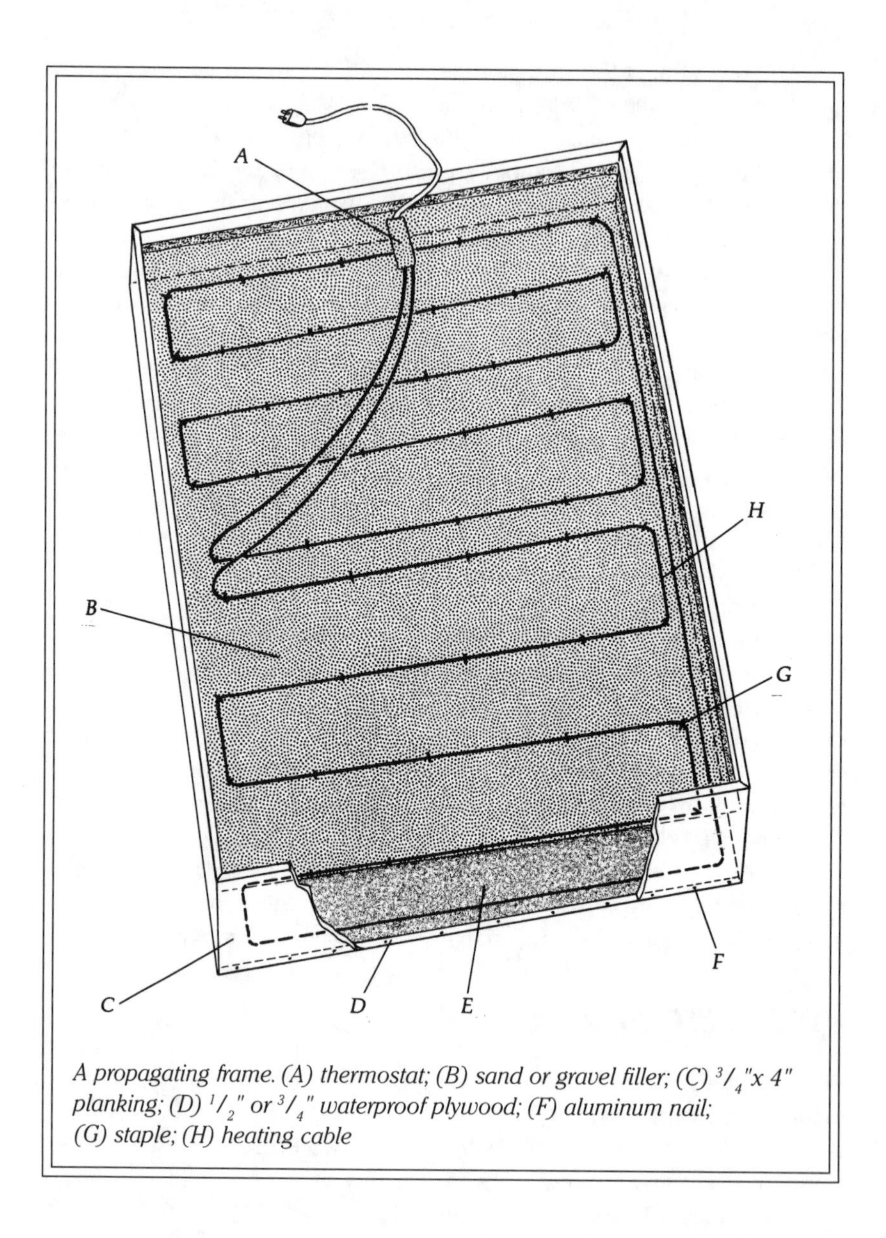

A propagating frame. (A) thermostat; (B) sand or gravel filler; (C) $^3/_4$"x 4" planking; (D) $^1/_2$" or $^3/_4$" waterproof plywood; (F) aluminum nail; (G) staple; (H) heating cable

CONTAINERS

Our present world suffers from a too much of muchness: too many cars, too many deodorants, too many can openers, and too many types of equipment for sowing seeds. Some, I admit, are a bit more aesthetic than old margarine cups, coffee cans, foil pans from frozen pies or TV dinners, and empty milk cartons—all of which make excellent seed flats. The new equipment also costs a good deal more than using clean food packaging left over from last night's dinner. Ev and Bruce Whittemore (see page 207) use Styrofoam coffee cups and have never purchased any other type of container.

So if you have some spare cash and a thirst for the new, the following containers are available from garden centers and nursery suppliers:

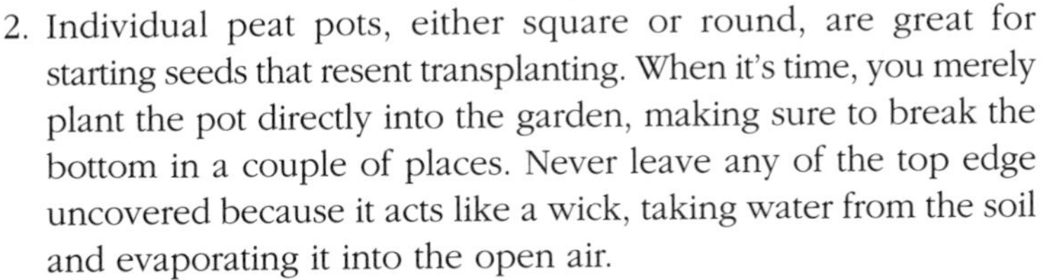
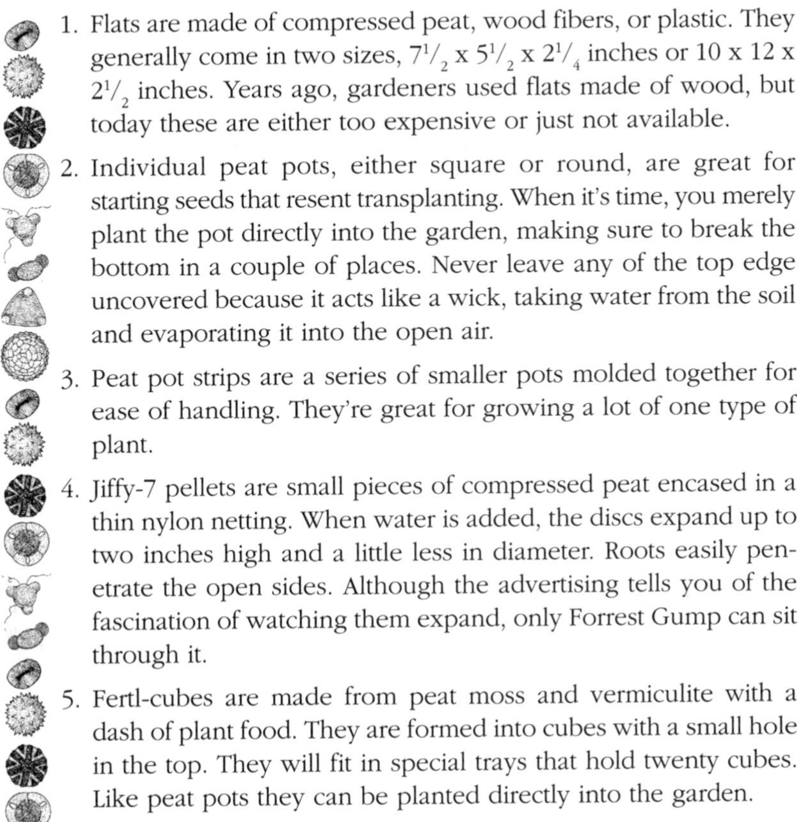

1. Flats are made of compressed peat, wood fibers, or plastic. They generally come in two sizes, $7^{1}/_{2}$ x $5^{1}/_{2}$ x $2^{1}/_{4}$ inches or 10 x 12 x $2^{1}/_{2}$ inches. Years ago, gardeners used flats made of wood, but today these are either too expensive or just not available.
2. Individual peat pots, either square or round, are great for starting seeds that resent transplanting. When it's time, you merely plant the pot directly into the garden, making sure to break the bottom in a couple of places. Never leave any of the top edge uncovered because it acts like a wick, taking water from the soil and evaporating it into the open air.
3. Peat pot strips are a series of smaller pots molded together for ease of handling. They're great for growing a lot of one type of plant.
4. Jiffy-7 pellets are small pieces of compressed peat encased in a thin nylon netting. When water is added, the discs expand up to two inches high and a little less in diameter. Roots easily penetrate the open sides. Although the advertising tells you of the fascination of watching them expand, only Forrest Gump can sit through it.
5. Fertl-cubes are made from peat moss and vermiculite with a dash of plant food. They are formed into cubes with a small hole in the top. They will fit in special trays that hold twenty cubes. Like peat pots they can be planted directly into the garden.

In addition, there are various combinations of the previous items, some made from compressed papers that expand into hexagonal containers, some with self-watering trays and wicks, all marketed as complete kits for seed propagation.

Recently, blocks of expanded polystyrene have appeared on the market. They measure about 6 x 3 x $2^{1}/_{2}$ inches and are perforated with eighteen holes that hold cylinders of compressed peat. They can be used again and again by simply replacing the starting cylinders.

Among the newer items are small wooden forms made of polished maple that form strips of newspaper into perfect little pots that are not only cheap, but also clean. Frankly, in our house, I'm now making my own paper pots and not using peat pots at all.

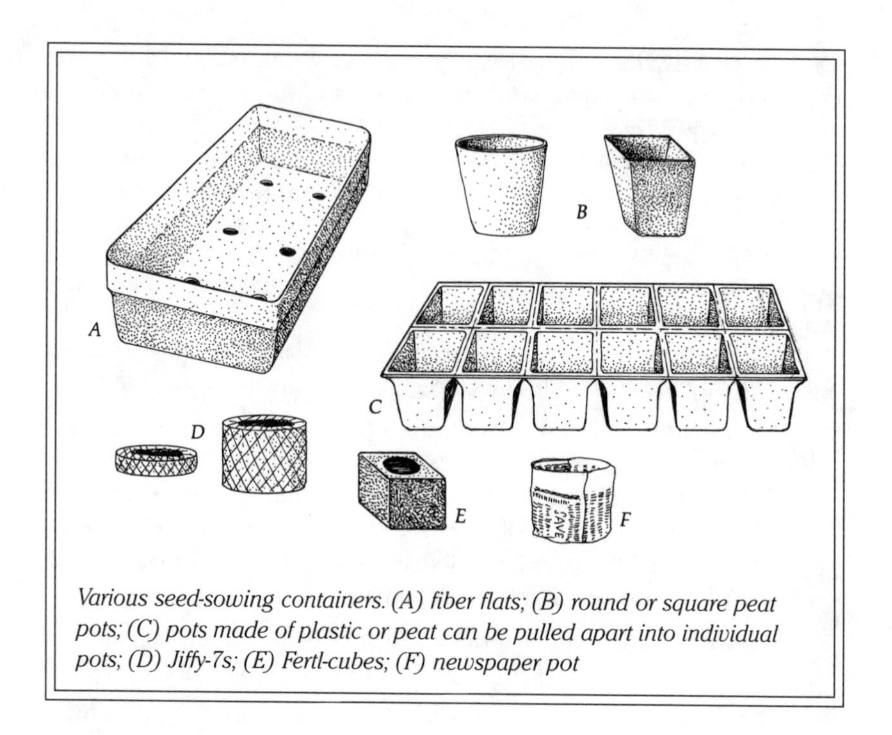

Various seed-sowing containers. (A) fiber flats; (B) round or square peat pots; (C) pots made of plastic or peat can be pulled apart into individual pots; (D) Jiffy-7s; (E) Fertl-cubes; (F) newspaper pot

There are also all sorts of plastic containers—anywhere from an inch to three or four inches across—that are ideal for sowing seeds. Just remember, they must be thoroughly cleaned after every use.

Many stores now sell small indoor greenhouses made of plastic to insure the proper warmth and humidity. Although they are no doubt stylish and much better looking in a small indoor setup than old metal trays with odd-sized pots under a ceiling of Saran wrap, they really don't work any better.

ADDING THE MIX TO CONTAINERS

Before adding your growing mix to containers, make sure it is thoroughly moistened. I've found the best way to wet the mix is to put some in a large plastic bag, add some warm water, then knead it like dough, continually adding more warm water until it's wet throughout. Only then do you add it to your chosen containers. Don't worry about this mix being too wet; whatever water it cannot absorb will leak out through the drainage holes in the various containers.

Another way is to put your ingredients in a large bowl, add warm water, stir well, and squeeze out all the excess water with your hands.

Fill your containers close to the top, leaving about a quarter-inch or a bit less free of mix so that air will easily circulate. Make sure the surface of the mix is flat, and you are ready to sow seeds.

TYPES OF SEEDS

There are perennials, biennials, and annuals. Short- or long-lived perennials often take two seasons to flower. After flowering they will live anywhere from a few years, like California poppies, or for more than a hundred, like most species of peonies. Biennials only live two years with the first year spent in slow growth and flowering, then going to seed in the second year. Annuals complete their entire life cycle in one garden season.

When grown from seed, most perennial plants spend the first year producing roots, and they flower only in the second, although there are exceptions to the rule.

Annuals are divided into three categories based on a system formulated in England and inherited by American nurseries: hardy, half-hardy, and tender. Many gardeners in the warmer sections of the United States think the designation *half-hardy* is superfluous, but I've found the term is useful. In my region of western North Carolina, for example, we have long, cool, and rainy springs, where temperatures seem to hover between 35°F and 50°F for weeks on end, and April showers are more of a truism than a cliche.

Hardy annuals are plants that tolerate a reasonable degree of frost, and even in the colder parts of the country many of these seeds survive a winter outside and germinate in the spring. The alternate freezes and thaws of late February and March will not harm them and are often necessary for germination.

Half-hardy annuals are usually damaged, set back, or killed by continued exposure to frost. However, most will stand up to an occasional light freeze and are impervious to endless days of cool, wet weather—a common occurence in the English climate.

Tender annuals come from the warmer parts of the world and need warm soils for germination. They are immediately killed by frost.

COATED OR PELLETED SEEDS

Coated or pelleted seeds were thought to be a big deal in the home nursery industry. The idea was to coat small seeds with various materials like clay to make them easier to handle. But it seems that coated seeds take longer to germinate because the water must first penetrate the added coat and then the actual seed coat. And pelleted seeds obviously cost more. I think you can do as well by mixing sand with the seeds.

Seed Tapes and Seed Blankets

Seed tapes are made from paper and usually come in fifteen-foot strips. The seeds are actually impregnated in the special paper tapes. The results are perfectly straight rows, uniform spacing, no bother with transplanting, and obviously no time spent in thinning little plants. The strips can be cut at any point to allow for multiple-row planting or parts saved for a second sowing. A number of vegetables, including beets, broccoli, cabbage, carrots, and parsnips come in seed tapes; flowers include alyssum, asters, cosmos, marigolds, statice, and zinnias.

Seed carpets are five-foot-square mats of tear-resistant yet porous fabrics, impregnated with various types of seeds. One such carpet is dedicated to wildflowers and contains a blend of fifteen flowers, including Johnny jump-ups, purple coneflowers, and blanket flowers (or gaillardias). Once in place, the plant roots grow through the lightweight fabric, and because the mix has both annuals and perennials, the gardener gets to see results within one season. But this is not a panacea; there is still work involved because the soil beneath the carpet must be prepared. That means workable ground. You cannot set a plant blanket down on flat, burned clay and expect great results or, for that matter, any results at all. But these carpets are obviously time-savers and have a great attraction for weekend gardeners.

CHIPPING SEEDS FOR GERMINATION

Certain seeds must have their seed coat broken to allow water to enter and stimulate the embryo into germination. Many seeds of the various morning glories (*Convolvulaceae*) and a number of the pea species (*Leguminoisae*) fall into this category. Usually seed packages will recommend chipping or using a file, scraping the seed coat. Soaking the seeds in water (start out with warm water) for twenty-four hours or until they begin to swell will accomplish the same goal. Moonflowers (*Ipomoea alba*) and castor beans (*Ricinus communis*) come to mind.

Over the years I have used a sharp X-Acto knife or a jackknife blade to chip the seed coat, taking only enough away to obviously break through the outer surface and expose the seed's interior.

TIME YOUR SOWING

Whether you sow seeds in a greenhouse, on a sunporch, or outdoors under a cold or hot frame or if you live in a part of the country where spring frosts occur, you must time your sowings so seeds can germinate and grow to a plantable size in time for your last spring frost. In Asheville, our last frost is usually around April 20. It takes about five weeks for marigolds to go from seed to plantable garden size, so I sow marigold seeds about March.

SOWING SEEDS

If you lack instructions about how deep to plant the seeds, use the following guide to determine the depth at which seeds should be planted: For seeds 1/16 of an inch or larger, cover them with the thickness of one seed. If seeds require light for germination, do not cover them but merely settle them into the mix with a misting of warm water.

TECHNIQUES FOR SOWING TINY SEEDS

Where I garden in western North Carolina, one of the many large-leaved trees that dot the mountainsides, and especially at the road's edge, are the Chinese empress trees (*Paulownia tomemtosa*). Their leaves are often a foot across, and a seedling tree can reach a height of six to eight feet in one or two seasons of growth. Yet the seeds are so small that a minor breeze can blow them out of your palm. The Tasmanian eucalyptus (*Eucalyptus globulus*) can easily reach a height of ninety feet, but its seeds are about the size and color of pepper that is ground for your salad at a trendy restaurant. And the seeds of orchids and begonias are like dust. Just because a plant makes a big statement doesn't mean its seeds are large.

When it comes to life after germination, small seeds are very surprising. Initially, when they sprout, the resultant plants are obviously tiny, mere dots of green. But with the proper light and water, they quickly grow, usually outstripping their larger cousins. In nature, most of the small-seeded plants come from areas where the soil is moist, and the germination bed is shaded from the hot tropic sun.

Your germination bed should be as flat as possible, so tap the growing mix with a small, flat piece of wood, after it's been put into the flat or container.

If you trust to your fingers to sow small seed, the seeds will usually wind up being too close together for proper growing. Luckily, there are several other ways of scattering. First, mix the seed with a small quantity of sharp sand, putting the resultant seed and sand mixture into an open envelope. Then, holding the open edge an inch or two above the compost, slowly move it back and forth across the surface, tapping the envelope gently as you go. Or you can again mix the seed with a small quantity of clean sharp sand, then use a folded piece of paper, again tapping it gently as you move it just above the surface of the planting mix. With either method, you can easily see where you sow because the light-colored sand will stand out against the darker mix.

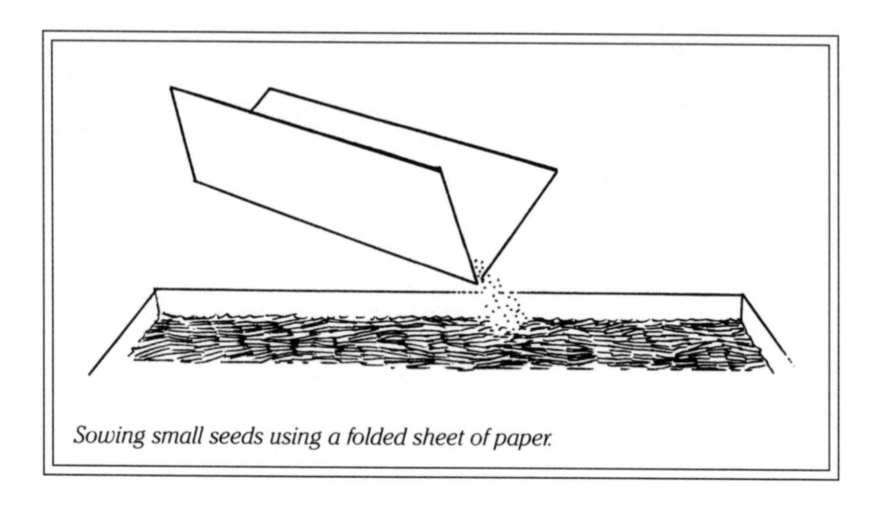

Sowing small seeds using a folded sheet of paper.

Another great trick to sowing small seeds involves adding Knox unflavored gelatin to the seed. I first heard about this trick from garden legends Doc and Katy Abraham of *The Green Thumb*. The gelatin gives each seedling an extra boost by providing both water and protein where it's needed. And because that protein is up to 16 percent nitrogen by weight, you are beginning the seedling's life with a mild fertilizer. The Abrahams suggest using a clean, dry saltshaker to sprinkle the mix over the medium.

Use a fine spray mister to settle the seeds into the mix, and do not cover. Until the seedlings are large enough to withstand the force of falling water, either continue to use the fine spray mister or set the flats into a larger pan of water, and let the mix soak up moisture from below.

COVERING THE CONTAINERS

Once the seeds are planted, add labels, then put the containers in a tray or plastic pan. If the container is deep enough, cover it with a sheet of glass or plastic. If using plastic wrap to cover the containers, use a plant label or small stick to act as a tent pole, so when drops of moisture condense, the plastic will not be weighted down and come to rest on the mix. These miniature greenhouses will prevent the medium from drying out.

Keep the various containers in your propagation box or in a warm spot away from the direct rays of the sun. Soon they will want all the sunlight you can provide, but not until they sprout. As soon as germination begins, bring the containers into bright light so the seedlings will not be weak and spindly. But remember, they must be shaded from the direct sun. They are still so small, they can quickly dry up and shrivel.

STARTING SEEDS UNDER LIGHTS

Depending on your discretionary income, various light fixtures are available to help in seed germination. These are especially helpful if you live in an apartment or have little room for plants. Even a dark basement can become a hothouse of seedlings.

Light fixtures range from small ones that use one or two wide-spectrum fluorescent tubes that fit on a tabletop to open structures with three or four tiers of shelves, each with multiple lights and capable of holding hundreds of seedlings. If your plant room is cool, polyethylene sheets can be draped over the lights to keep the heat trapped and maintain the necessary temperatures to insure germination.

Recently, an apparatus called The Lighthouse has appeared on the market. It consists of heavy cardboard coated on one side with aluminum foil. When the apparatus is left in a window, the foil reflects sunlight around the seedlings, leading to strong, well-formed plants.

SOWING SEEDS OUTDOORS

Many seeds do not need the protected environment discussed thus far. They can be sowed directly into your garden. Of course, the soil must be properly prepared and raked so it's level. Following the directions on the seed packet is the best starting point, but it's useful to remember the following rules:

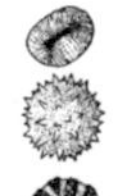

1. Never sow seed onto cool, wet soil. Wet soil crusts over when it dries, and seeds find it difficult to break through with tender roots. Many seeds tend to rot when confronted with cold, wet soil.
2. Never plant the seed too deeply in the soil. It's always better to err on the shallow side. It's amazing how fast roots can grow down but how often a too-heavy layer of soil prevents the germinating seed from growing up.
3. Unless random gardening is your aim, sow seeds in rows or circles. Often, emergent seedlings will look exactly like weeds, especially those of the ornamental grasses. When planted in rows, the valuable seedlings are easy to spot.

ONCE THE SEEDLINGS APPEAR

When the first green shoots appear, move the containers into bright light. Then, after a few days of acclimation, move them again into the sunlight in a greenhouse or to a window. Once the spring season is past, use a protection such as screening during the very hot rays of the noonday sun. Turn the seedlings daily to keep them from bending to the light. If you have time every day to check on water requirements, you can remove the plastic covers.

If my seedlings are in the greenhouse, I always connect a small fan so there is constant air circulation above the seedlings. This is just in case there might be some fungus spores about. The drying effect of the moving air prevents the spores from causing problems.

WATERING SEEDLINGS

When the mix begins to dry, either water gently from the bottom through the drainage holes or from the top, making sure to use a gentle mister or a fine-holed rose. One of those rubber bulbs used to moisten clothes before ironing (if anybody still irons) is perfect for watering seedlings. Just be sure to use warm water.

If using large peat flats for seedlings, you can water them from below by making a hole in the bottom of the pan. Insert a cotton wick—similar to those used in kerosene lamps—through the hole and then into a pan of water kept below.

Once the true leaves appear, the seedlings have just about used up their food reserves. If you have used soil or any other prepared mediums that contain nutrients, you can get by with just warm water. But if you have used a nutrient-sterile medium, now's the time to use a liquid fertilizer and make up for the lack of nutrition in the soil. Follow mixing instructions, then dilute the solution by one-half. Even if you are using soil, it never hurts to give some extra food, so I always fertilize all the seedlings.

FIVE RULES FOR PROPER GERMINATION

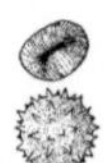

1. Don't overwater. It deprives the seed of needed oxygen.
2. Never sow seeds too deeply. If in doubt, just cover them lightly.

3. Never let the mix dry out.

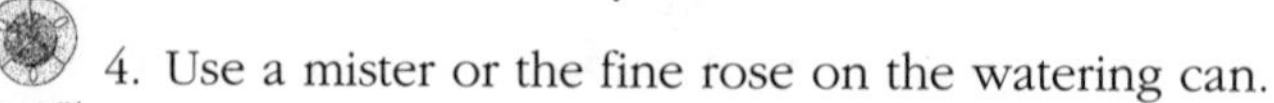

4. Use a mister or the fine rose on the watering can.

5. Keep seeds away from sunlight until germination is complete.

CONTROLLING INSECTS

If very small seedlings are attacked by an insect horde such as an army of aphids, I steal a march on these loathsome pests and remove them one by one with a toothpick. For whitefly, I hang pieces of yellow, glued paper nearby. These pest strips are now available from nursery suppliers. When in place, I fan the tiny flies toward their doom. The only other minute pests are spider mites, almost invisible pests that work on the undersides of leaves, spinning thin webs and sucking the life out of leaves. These arachnids can be removed with sprays of warm water. If seedlings have become young plants and the pests have multiplied, I use one of the commercially available cleaning soaps that come in spray bottles. First, spray the beasts, then quickly wash away the soap with sprays of warm water.

FUNGUS GNATS

Fungus gnats are tiny, black, mosquitolike flying insects that persist whenever peat moss or peat products are used. They eat organic matter and lay eggs in soil or any organic medium used for plants. Although minute, they are a pest, and I'm firmly convinced that the larvae (teeny, threadlike white worms), when they occur in large numbers, will feed on seedling and plant roots, causing a lack of vigor in larger plants and irreparable damage with little seedlings.

There are larvicides (a fancy name for chemical-attacking insect larvae) that can be mixed with water and used as a drench to control these pests. They are available from commercial horticultural sources. But the best way to fight these pests is to use absolutely clean and sterile pots, containers, and soil mixes.

TRANSPLANTING SEEDS

If seeds have been planted as individuals in their own pots, you need only repot when they have outgrown their original homestead. For other seedlings, thin them out to at least one inch between plants. They then can be left alone until large enough and easily handled.

If you want to separate many closely grown seedlings, make sure the soil is moist, then spoon out a bunch and carefully pry the plants apart with your fingers, trying not to break any of the roots. In fact, always handle a seedling by its leaves, never by the stem or the roots.

Make holes in a new soil-filled container, using a pencil or a wooden dowel. Drop in the seedlings, easing the roots down into the hole. Then with that pencil, push the soil around the roots and gently settle roots in with warm water. Keep the seedlings about an inch apart.

The pointed end of a knife or a plastic plant label makes an excellent tool for transplanting. Just be gentle! Pick up the tiny plant, move it to a new home like a larger pot, then lightly cover the roots with damp soil, and with a little more water, wash in the roots.

Following a transplant, keep the plants in subdued light for several days. Then gradually increase light intensity.

GROWING ON

Plants are ready to be moved outdoors when the roots show through the bottom of the pot or, as in peat pots, begin to grow right through the sides. Some plants benefit from pinching when they get too tall in proportion to their total growth. Petunias, snapdragons, zinnias, ageratums, browallias, annual chrysanthemums, and annual phlox respond to a good pinch. The result is bushier plants with more flowers.

Don't pinch balsam, cockscomb, helicrysums, or poppies.

HARDENING OFF

When seedlings are large enough to be moved outdoors, they should go through a period called hardening off. Until now the leaves have been protected from the elements, but remember that surface cells are very thin and unable to protect the inner cells from the full force of sunlight. So we first move the seedlings to a shaded cold frame, a porch, or under a tree, in fact, any place where they are sheltered from the sun.

At the same time, keep the transplants off the chilly ground by setting them on a raised surface. This not only helps roots to acclimate, but also helps to prevent garden thugs, like slugs, from reaching tender leaves. If nights get too chilly, move the plants back inside, or cover them with any light material that will keep in the ground heat but will not smother or crush the plants.

After three days, give the transplants morning sun and afternoon shade. Also, begin to decrease the frequency of watering, and withhold fertilizing at this time.

After three more days, give the transplants all-day sun. After another week, they can take cooler nights. Don't forget that hardy annuals will take lots of frosts; half-hardy annuals will only live through light frosts; and tender annuals cannot take any frost at all.

The Importance of a Cold Frame

Gardeners, especially those who live in a climate with late spring frosts, can use an old concept called the cold frame. Cold frames can be fancy or plain but consist of a small garden area that is roofed over with glass and has sides protected from strong winds. Once spring is on its way and the earth begins to warm, seedlings and transplants can be left out in the open by day, but if a freeze is predicted, the plants in the cold frame will be covered over and protected from the cold. If the sun is too bright, screening can be put in place for additional protection. Being in the open air is the best thing to toughen seedlings for the life ahead.

THE IMPORTANCE OF SHADE

Blistering sun can be a benefit to an old yucca (*Yucca* spp.) or a mature hedge of boxwood (*Buxus* spp.), but it usually spells death for the majority of seedling plants. Usually in nature, they are sheltered by taller plants or grasses, giving them a chance to become established before that hot sun hits them with all its power. That means that new seedlings in your garden also need protection.

Many catalogs now carry a plastic shade cloth that can be spread out on metal hoops or even used like a small tent. We use many old window screens, held over the plants by concrete blocks placed at corners. Very small plants can be protected by a fiber seed flat propped up by a stick or even by newspaper tents or an old coffee can.

TRANSPLANTING TO LARGER CONTAINERS

If transplants are not taken directly to the garden, they must be moved on to larger pots or to a nursery bed. If moved to a pot, the transplant will get quality soil and a chance to grow roots under your supervision. If you use peat pots, you can plant the pot, the plant, and the roots with ease. Or if you use small plastic pots, transplants are easily removed, and the roots will hold most of the soil together while they go into their new hole. Plants with tap roots should go immediately to their permanent site.

It's a good idea to perform these operations on cool and cloudy days to cut down on stress.

A NURSERY BED

A nursery bed is a small area of protected good-quality earth where young seedlings get a chance to grow without running into too much sun, too much cold, and too little water. Any gardener who intends raising a number of plants from seed should plan a nursery bed, close enough to the house so it will not be easily ignored and will have plenty of nearby water.

The dried, papery flowers of pearly everlasting (Anaphalis margaritacea) *send tiny bristly seeds on the wind.*

CHAPTER 16

Success with Seeds

THE WHITTEMORE ALPINE GARDEN

Nobody benefits from seeds more than rock gardeners. Without the various seed exchanges sponsored by all those associations that delight in what are termed *alpine* or *rock garden plants*, the choices for rock gardeners would be few and mostly uninspired. But once you belong, the world of the unusual becomes as close as your mailbox.

One such garden belongs to Ev and Bruce Whittemore. They call their garden Fort Courage. It's an almost-hilltop garden at the end of a twisting up-mountain road about eight miles outside of Brevard, North Carolina.

Eventually, the garden will cover three acres. There are areas of scree where the Whittemores have moved tons of crushed rock by the individual pail in much the same manner that Chinese peasants moved farm materials before the collective revolution. All this stone was needed to provide necessary drainage for growing and keeping alpine plants. There are conifer beds, neat and orderly seed frames, and areas of naturalized wildflowers and lilies. There is a path that leads to a real cliff that started out looking like a rhododendron and pre-president Mount Rushmore but now has that controlled look of nature when it's guided by the hand of man.

It's true. Everywhere you look in this garden, you see the gardener's hand—and to anyone who has ever gardened on a mountaintop where the rock is plentiful and level areas are few—it's evident that a massive amount of physical labor has passed this way.

Ev and Bruce look trim, and their eyes are bright with that sense of accomplishment that comes from a project well done. They took a few moments away from the construction of a new seed flat and told me some of the garden's history.

"Bruce had seven more years before retiring, and I told him I wasn't going to make it," said Ev, stooping to pick up a weed and brushing the hair from her eyes.

> *We then lived on a half-acre with a greenhouse stuffed with plants and a finished garden with no room for any additions. After checking weather statistics around the country, the mountains of North Carolina turned out to be the answer. A further search found us this five acres that we immediately dubbed Fort Courage.*

In April of 1983, Bruce and Ev took a three-week vacation and lived on their new property using a borrowed tent. They began construction on an eight-by-ten-foot building, ultimately meant for a garden shed, but until the final move, it would be Ev's home during growing seasons over the next few years.

"There was no electricity," she said, "and all the wood had to be cut by hand. It was our first and last time living in a tent. There were nights of record cold, and we had an inch of snow one morning."

The next year they built an eight-by-ten-foot addition to the main garden shed. This new part of the house had enough room for a small chest of drawers and a potting bench. They added a tiny deck so they could have their meals outside while listening to the magic of the tumbling stream that ran nearby.

Today, it's easy to walk the pathways of the garden, but Ev quickly points out, "All the garden areas had to be chain-sawed and cleared of mature trees and twenty- to thirty-foot rhododendrons, including *Rhododendron maximum* and *Kalmia latifolia*. We actually had to burn the limbs and trunks."

> *Many of the plants have been started from seed that we swapped with other enthusiasts throughout the world. I find gardeners in Czechoslovakia appreciate subscriptions to our gardening magazines, and even empty Windex bottles are appreciated in return for seeds. I traded seeds with a friend in Japan who grows his plants in pots sunk in the ground, and on occasion we journey to the West to collect more.*
>
> *I like a natural look in the garden, and this is created by using several plants of each kind mixed with a few single specimens. A collection of just one plant of each hybrid or species isn't my idea of a garden. One never sees one* Draba densifolia, *one* Anemone patens, *one* Silene acaulis, *and one* Arenaria obtusiloba *in the mountains; they grow in drifts. But unless you have deep pockets, you do not buy drifts of each. Growing from seed can give you the privilege of numbers*

and the ability to achieve a natural effect in your garden, without major expense.

No matter how hard you work at selecting the best conditions for growing unusual plants in the garden, trough, or alpine house, there is never a guarantee that any of them will survive. The worst part of having to watch a choice plant fail, then die, is watching it leave this world in slow motion. Ev hedges her bets by growing from seed, then placing seedlings in several different garden spots. She also notes that extra plants grown from seed can be used for barter or gifts among gardeners—and she speaks from experience. Many visitors to the Whittemore gardens bring favorite plants along to test Ev's identification abilities, then leave the new plant but take another home in its stead.

Their day begins with a drive to nearby Brevard for breakfast at one of three fast-food restaurants. They usually eat a biscuit with jelly and have a cup of black coffee, do a few errands, and head back for a full day in the garden.

One of the perks of fast food is the marvelous, tough, insulated cups used to serve coffee. The Whittemores collect every cup they use, rinse it well, then burn drainage holes in the bottoms, using a small electric soldering iron. These become the seed pots for the next generation of plants.

I rarely waste the time and expense of labeling a seed pot until germination has occurred. The cups take a Sharpie fine point pen beautifully, and there is lots of room for germination information on the cup. This is easily transferred to a label if the seeds germinate. If I run out of cups—and we drink a lot of coffee—I turn to four-inch square pots. I now use larger seed containers than I once did, having realized that young seedlings will hold over in better form in larger pots if I'm busy and can't pot up when I should.

Their potting mix is uncomplicated and not very scientific. There are always bags of commercial potting soil in the garden shed, and one bag is unloaded in the wheelbarrow. Since gritty creek sand is purchased by the truckload for drainage in constructing a new garden or sand molds for troughs, there is always some in the yard. A few shovels of sand are added to the soil. If ericaceous seeds are being sown, damp peat moss will be run through a sifter, then added. If scree plants are the order of the day, the peat is omitted and a pail of turkey grit is added. Ev buys the grit from the feed-and-seed stores in eighty-pound bags. After the seed pots are filled, she shovels any leftover mix into a large plastic container for future use.

"But there are days when I mix a bag of commercial perlite-and-peat with the pail of turkey grit. I try for as sterile a mix as possible to avoid weed seeds, but I don't get uptight if sterility isn't perfect."

Ev also suggests that no large pieces of peat or wood chips are left in your mix. Trying to separate the roots of sour small seedlings all determined to hang on to a wooden shard will result in one unhappy gardener—and four unhappy seedlings.

Their favorite time for sowing is mid-December, but they have been known to plant seeds all year long. Seeds waiting for the pot are kept in a covered cookie tin in the crisper drawer of the refrigerator.

When sowing, the pots are filled with soil and firmed by being banged sharply on a hard surface. The soil mix comes close to the top, leaving just enough room for a thin mulch. This allows for plenty of air movement between the seedlings. Seeds are sprinkled on top and covered lightly with turkey grit. The pots are then set in water to soak overnight, watered from the top with a watering can, and drained.

With December sowing, the pots are set in the laundry for three weeks. Whatever germinates goes under the lighting system or on the window sill to grow, and the other reluctant seed pots are put into the outside frame. Lengths of screening are placed on top to protect the pots from the family poodle, who finds them a comfortable place for a nap during the heavy mountain rainstorms. If seeds are sown in warm weather, the period in the laundry is omitted, and the pots go directly out to the frame. When they germinate, pots are sunk in a sand frame with no screening.

According to Ev, anyone can build a frame. She considers frames essential to gardening, and sand frames are her favorite. Preserved or treated timbers are easy to lay after leveling the ground. If they are only two or three timbers high, she omits the nails. Ev once made a three-by-six-foot frame of concrete blocks but hated it, so she spread hypertuffa on the outside and turned it into a large trough. Bricks would also make a good frame but wouldn't fit within the Whittemore garden because nothing else in the garden is constructed of brick.

When constructing a simple frame of concrete blocks, she would use two or three layers in a rectangular or square configuration. Inside, sand would be added almost to the top and pots were sunk in this medium to about a half-inch from the pot's top. After the sand is wet, it's easy to lift a pot, check it, and then return it to the same hole. Sawdust can be used as a substitute, but sand is her first choice.

"The sun can be brutal in North Carolina," she said. "Using mulched, insulated cups sunk in a sand frame protects the roots of tiny seedlings from heat and cuts down on watering. I prefer growing my plants hard, and plants not meant for shade gardens are given full sun while shade-lovers get treated to shade screens."

Then there is the latest addition to Fort Courage: An alpine house for mountain plants that need a cold winter but come from areas where the snow begins to fall in September and stays unmelted until March or April. Building the alpine house is a story in itself.

The Whittemores got the idea for an alpine house last summer while visiting Rick Luff, a friend living in Graham, Washington. Rick told them the house cost $250 and was finished in a weekend, so the Whittemores decided on a ten-by-fourteen-foot house.

Their first problem arose with the site. Rick's structure went up on level ground. "It took six days of mattocking and digging in clay," said Ev, "to convince us the job was not going to be finished in one weekend."

Finally the clay was removed, but to level the floor, they needed 135 concrete blocks. Bringing concrete blocks to Fort Courage is not as simple as picking up the phone and having them delivered and stacked in the backyard.

"After the foundation was completed," said Ev, "I filled the holes in the blocks with grit to keep out any little visitors. Next, we laid pressurized four-by-fours for a base and installed eight twenty-foot PVC pipes as a framework for the poly covering, and Bruce framed in each end with more treated wood."

With a wheel-barrow, Ev moved about seven tons of crushed fine rock, most of it used for the flooring and around the nearby patio. In addition, five cubic yards of creek sand sat in the driveway, and as soon as Bruce finished the electrical work, it had to be moved—again by hand.

Bruce's jobs included electrical connections: a light over the door that can be turned on from the garage, two fans necessary for air movement, and an exhaust fan in the center peak of the roof that activates when the temperature up there reaches 75°F. In addition, there are plugs for portable heaters. To accomplish the wiring, Bruce had to break through the back of the garage wall and bring in a line from the main house.

"After moving in the creek sand," continued Ev, "I mixed my medium for growing the plants, and I mixed lots of it. A ten-by-fourteen-foot area takes a lot of fill. The project took a month to complete, and needless to say, with a foundation, poly pipes, treated wood, poly covering, Plexiglas, 135 concrete blocks, automatic vents, exhaust fan with casing, electrical material, circulating fans, pressurized timbers, not to mention labor, $250 was a low estimate."

By September 19, the alpine house was finished. Ev adjusted her glasses and looked down at some six hundred Styrofoam cups, all with individual seedlings of rare and unusual plants. "My original ten-year plan," she said, "called for seven years to do most of the hard labor, and we often worked ten to twelve hours a day. But we've cut down a bit since then, and now a typical day in the garden is six to eight hours."

After hearing the story behind the garden and the hours of work logged in its completion, we looked about the pathways lined with blooming plants all waving their colors under a mountain sky and marveled at a job well done. "Good Lord," was all we could say.

What was wilderness is now a world of plants. The left side of the driveway, for example, has a fifty-foot wall of rock all planted with lewisias, *Penstemon rupicola, Dryas octopetala minor, Thymus caespititus* in the sun, and *Gymnocarpum dryopteris plumosum* and *Hedera helix* "Duckfoot" in the shade. Everything is mulched with sawdust, itself covered with pine needles for a more attractive appearance.

And every morning about half past ten, a breeze comes sweeping up the driveway, but the sawdust is held in place by the needles.

But for all the seeds that do germinate, there are many that do not. Ev recommends that beginners start small by sowing about a dozen different types of fairly easy-to-germinate seeds. Most will germinate and give beginners encouragement to do more.

Germinating seeds requires a certain amount of commitment. Pots must be watered during dry spells; growing seedlings must be moved to larger pots. Then there is that last step of planting seedlings out in the garden. According to Ev:

> *Members of the American Rock Garden Society who are mad about growing from seeds have their favorite private or plant society sources, some being carefully guarded. Seed collectors advertise in society bulletins and if I were a beginner and wanted to get started growing plants from seed, I'd go to local chapter meetings and visit area rock gardens to see what's being successfully grown and what would appeal to me. Talking with gardeners can give you many clues to seed sources and growing conditions. Rock gardeners are usually generous with their time and cultural hints.*

When looking through society seed lists, Ev pointed out that when more than three donors send in the same seed, it could be an easy plant to grow. She suggests the following for a beginner: *Androsace carnea, Aquilegia canadensis* "Nana" or *A. fllabellata, Aster alpinus, Campanula carpatica,* any *Dianthus* species, *Draba aizoides, Erigeron compositus, Erinus alpinus, Gentiana septemfida, Papaver burseri, Potentilla megalantha,* and *Viola jooi.*

According to Ev, seed sowing is addictive, like eating chocolate chip cookies or potato chips. Start with a small seed assortment, a small frame, and a small garden. Eventually, you will be the proud owner of a garden filled with rare treasures started from seed.

CHAPTER 17

Propagating Ferns

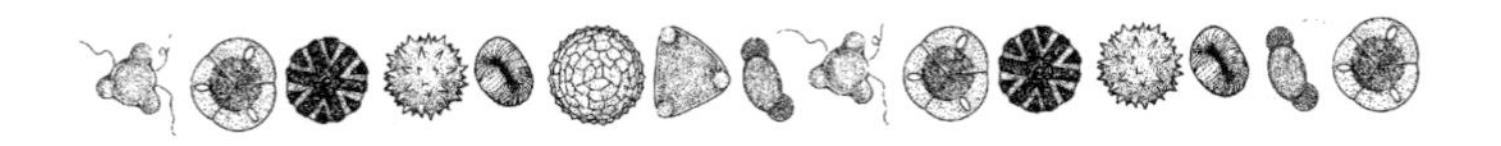

It was the 1960s, and our first apartment was a floor-through in Brooklyn, about three blocks from Grand Army Plaza. My wife and I came from a small village in upstate New York, and though smitten by Gordon Jenkins's *Manhattan Tower*, we were willing to trade just a bit of sophistication for a touch of home—and that included the color green. The apartment's sunny rear windows opened on a massive fire escape laced with grills, locks, and blinds—not a setup for plants. The front windows, however, while lacking sunlight, had class: Wavy glass sat within deep frames, flanked by those marvelous folding and louvered shutters found in old brownstones. It seemed a perfect place to start a small indoor plant collection. But what to grow?

About two weeks later I was walking down Fifty-third Street and passed a small shop with a number of ferns displayed in hanging baskets. In the center of the group was a magnificent crisped blue fern (*Polypodium aureum* "Mandaianum"). Back in the sixties, instant purchases were still possible, and I bought the plant on the spot.

Within a year I had learned a great deal about fern culture, including the fact that the fern leaf is called the blade or frond, and the stem is referred to as the stalk or stipe. And spring's newly emerged fern stalks are called fiddleheads, or croziers, because the outline closely resembles that of the fiddle's top or a bishop's staff. Fronds, I discovered, vary from simple undivided leaves to those with compound outlines that are typically lacy, resembling the ferns that most people know. The complicated divisions are composed of individual leaflets and subdivided leaflets, or pinnules, and can be divided again into lobes or pinnulets. But whatever the nomenclature, these were all marvelous plants, and our indoor window garden soon contained a number of classic ferns.

PROPAGATING FERNS

In Shakespeare's *Henry IV*, Gadshill, a crony of Sir John Falstaff, explains to the court chamberlain: "We have the receipt of fern-seed, we walk invisible." To which the chamberlain replies: "Nay, by my faith, I think you are more beholding to the night than to fern-seed for your walking invisible."

The good chamberlain (and I assume Shakespeare) expresses a healthy skepticism about this particular magic, but it was not until the early 1800s that the mechanism of fern reproduction was finally resolved and the true fern seed discovered.

Ferns do not bear flowers and seeds but begin their life cycle by producing spores. Spores are reproductive bodies formed by ferns and algae, lichens, fungi, and mosses as well. They usually contain but one cell and are only the first of many steps leading to a new plant (see page 5). Spores are contained within clusters of spore cases called sporangia. In turn, the sporangia are clustered in an individual patche called a sorus, the plural being sori. Look for sori on the undersides of fern leaves where they are often mistaken for insect infestations or disease by wary houseplant owners.

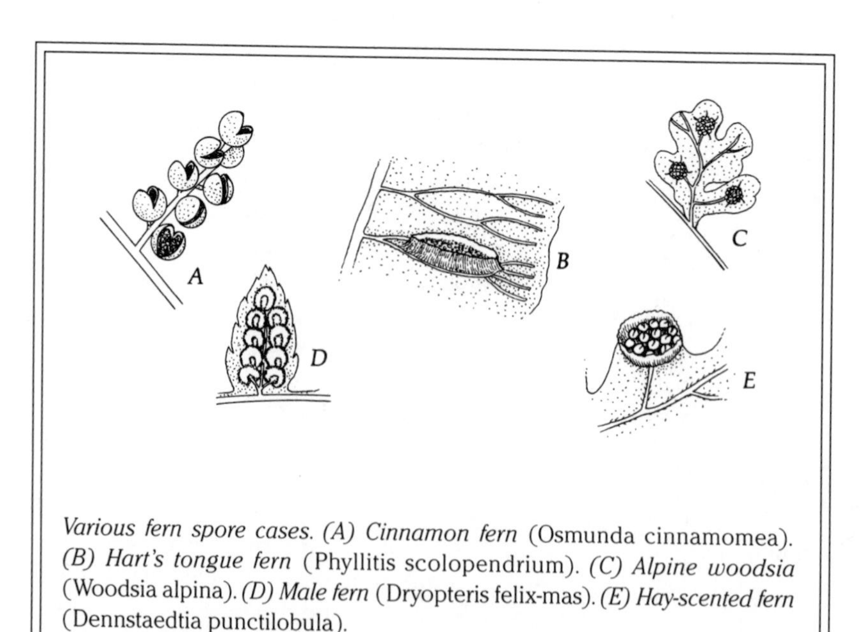

Various fern spore cases. (A) Cinnamon fern (Osmunda cinnamomea). *(B) Hart's tongue fern* (Phyllitis scolopendrium). *(C) Alpine woodsia* (Woodsia alpina). *(D) Male fern* (Dryopteris felix-mas). *(E) Hay-scented fern* (Dennstaedtia punctilobula).

Sori come in a wide variety of shapes and sizes such as round, oblong, curved or kidney shaped, or sometimes in a straight line; and because the shapes vary according to the species, they are used by botanists for identification.

Sporangia are at first green but turn brown with age. When mature—and weather conditions are fairly dry—the walls of the sporangia bend back and usually discharge their dustlike spores into the open air. Because there are dozens of spore cases in every sorus and because each leaflet will bear up to twelve sori with hundreds of leaflets on each fertile leaf and a number of such leaves in every plant, the total spore count is easily in the billions.

If it's raining, the sporangia remain closed, but if the day is fair and the humidity is low, the spores fly into the open air. They can either travel for miles on a breeze or fall within inches of the parent plant. Then, with proper temperature and adequate moisture, each spore develops into a flat, heart-shaped structure called a prothallium. When mature, prothallia are about the size of an aspirin tablet and bear no similarity to a traditional fern frond.

Separate male and female organs grow in the underside of the prothallia and produce eggs and sperm. With water supplied by dew, rain, fog, or even melting snow, the sperms swim to the egg, and fertilization occurs. In a short time, an embryo fern begins to grow.

Because the growing conditions needed by fern spores match those of mold, funguses, and a host of other undesirables, absolutely sterile conditions are needed when propagating ferns with these methods.

I use three- or four-inch clay pots that have been scrubbed with a fingernail brush and then boiled. The pots are crocked and filled to one inch of the rim with a sterilized soil mix of one-third potting soil, one-third peat moss, and one-third sharp sand. If unsure about the sterility of the mix, bake the soil in a 250°F oven for two hours. Although it all sounds like a hospital soap opera on TV, there's no sense in setting everything up to be done in by a wandering fungus or a clambering slime mold.

Now soak the pots and mix with boiled, then cooled, tap water until they are completely saturated. Drain off the excess water. Cast the spores one species to a pot, and label carefully because baby ferns all look alike. Cover the pots with a sheet of glass, unbending plastic, or a clinging plastic wrap, and keep them in a warm spot of 65°F–70°F (18°C–21°C), in dim light. Keep pots away from direct sunlight, and never let them dry out, always watering from the bottom so germinating spores are not disturbed. If condensation on the glass or plastic becomes too heavy, remove the cover for a time.

Many species will germinate within a few days; others take a few weeks. As the prothallia grow, a green cast will appear on the surface, and in about three months, they should reach their full size. If spore germination has been very good, thin out the prothallia so they are about one inch apart. Falling condensation from the overhead cover should provide enough water for the sperms to swim to the eggs, but you can mist the surface if the mix shows signs of drying out, remembering to use boiled, then cooled, tap water.

Tiny ferns will now develop, growing from the center of the prothallium. As the fronds grow larger, you'll have to thin the plants again. When three or more fronds have appeared, transplant to individual pots. Two years will be needed for a mature plant.

When ferns grow in a terrarium, baby ferns will often pop up in unlikely places. Some years ago, we kept a mango tree in a large self-watering pot. One morning I noticed that the soil was covered with green. When I examined it with a hand lens, I found hundreds of prothallia that came from the spores of a fan table fern (*Pteris cretica* "Wilsonii") located about five pots away.

CHAPTER 18

Collecting Your Own Seed

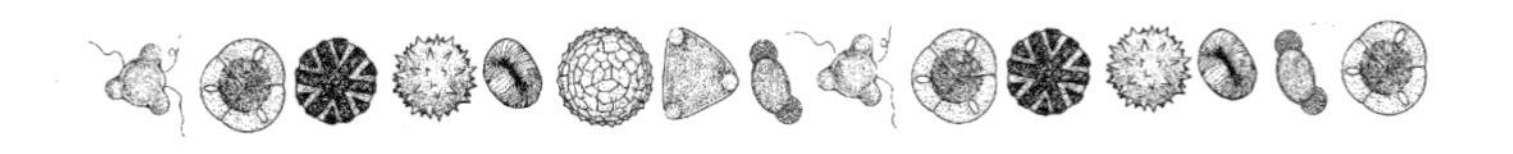

Until I joined the American Rock Garden Society and was introduced to a number of gardeners across the country, I never collected seed. I was far too busy and impatient to bother keeping seed of what, because it grew in my garden, I thought was common and old hat. I knew my wife saved all types of vegetable seeds, but I've never been big on vegetables; I usually don't grow what anybody can eat. Then I found gardeners would come into my garden and see a particular species of, let's say, a certain *Potentilla*, and immediately asked where it came from. I had to answer that this particular plant came from a rock gardener friend by the name of Budd Myers, who in turn, had traded it with somebody else. Suddenly, I knew it was time to start saving seed.

Collecting your own wildflower seed can be a wonderful experience. Once the flowers have passed, even the most sophisticated gardener usually loses interest in what's left. When gathering your own seed, you begin to look closely at the method a particular plant has chosen, not only to produce seed, but also to spread it about. By collecting seeds and spores, you become acquainted with the incredible pincers of the seed pod of the unicorn plant (*Proboscidea louisianica*) or the long dark brown pods of the locust (*Robinia pseudocacia*) or the colorful clusters of deep red berries that mark the staghorn sumac (*Rhubus styphina*). After the petals fall from the starflower or star everlasting (*Scabiosa stellata*), a ball of dozens of florets remains behind, each with a star-shaped center, making this particular plant more attractive in seed indigo (*Baptisia australis*)?

Lily seeds are thin and flat, usually packed like slices of bread, then stored in brown pods that begin to split when the seed is ripe. Seeds of the Siberian pea shrub (*Caragana* spp.) are packed within a pod usually containing about six seeds. On the other hand, grass seeds number in the thousands, appearing in airy terminal spikes, panicles, and racemes, usually turning brown when ready for harvest. Poppy seeds come in their own shaker, and if the pods are not collected just before ripening, the wind will soon send the seeds out to the rest of the world like sale from a saltcellar.

And if you think the seeds and seed pods of some plants are truly beautiful when viewed life-size, wait until you examine some of them under a low-powered lens.

Last year when the flowers of *Hibiscus coccineus* were in full bloom, I noticed that one plant of six-foot stature, topped with such intense red flowers, needed like company to really make a statement. I had purchased the plant two years previously at Thomas Jefferson's Monticello from their marvelous plant shop, so it was not only beautiful, but also memorable.

The seed pods are a papery brown about one and a half inches across, and the seeds ripened by the end of September. But I neglected to harvest them, and many succumbed to being food for autumnal insects. By the time I arrived with a paper bag in hand, the collected seeds numbered eighteen.

I stored them in a box in the laundry room, where winter temperatures hover around 45°F, and last March, planted them in a sterile mix, topped with a thin layer of sphagnum moss.

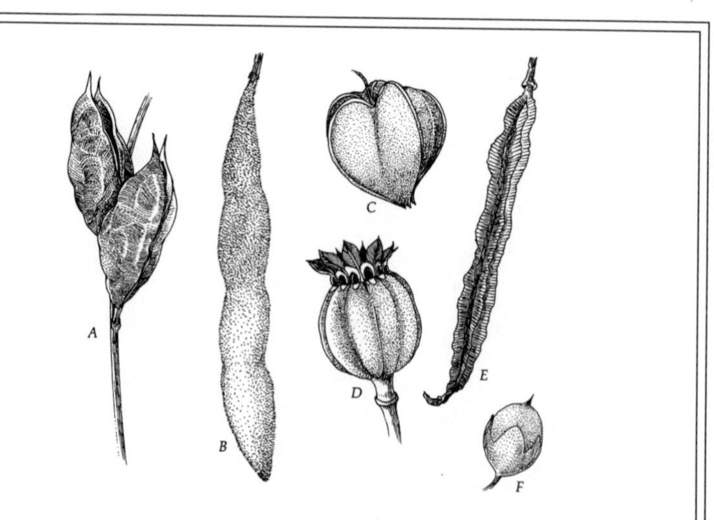

Various seed pods. (A) False indigo (Baptisia australis). *(B) Wisteria* (Wisteria sinensis). *(C) Husk tomato* (Physalis alkekengi). *(D) Poppy* (Papaver somniferum). *(E) Mamane* (Sophora chrysophylla). *(F) Flax* (Linum Perenne).

They germinated in mid-April, and recently, I transplanted fifteen healthy seedlings into fifteen little paper pots ready to go out to the nursery bed for the summer, and all set to bloom next year.

I grow a lot of hellebores (*Helleborus* spp.) in my garden. Up until a few years ago, most gardeners never knew about this great genus of plants, but today, their fame is spreading, and the best way to get new plants is by seed.

The seed pods are found ranged around the center of the flowers' remains. As soon as the pods begin to split, the flowers should be cut off and left in a dish or a paper bag until all the seed escapes. But there's a problem with these plants: The seed is not long-lived. The longer it's stored, the worse the germination, and contrary to most of the garden books, the seeds do not need stratification to germinate.

As soon as the seeds are available, I immediately sow three seeds in small paper pots, using sterile potting soil topped with a layer of sphagnum moss. The individual pots are then put in a plastic tray with a clear top and set out of the sun in a cool spot on the back deck. When they germinate, which usually begins by fall, I pick the best of the three. The seedlings spend the winter in a cold greenhouse and are then potted up the following spring.

As described, with two different plants, there are two different methods of dealing with collected seeds. By checking the books in the bibiliography, by joining the various seed societies mentioned in chapter 13, and by asking garden friends and neighbors, your knowledge should soon be formidable.

As to the actual seed collecting, Thoreau once said, "Beware of all enterprises that require new clothes," and, of course, he knew of what he wrote. The same holds true for most gardening activites. Once you have a good trowel and shovel, pants well constructed at the knees, and a good hat, you can get by with little else. The same holds true with seed collecting.

Commercial seed nurseries use raking and gathering tools called seed strippers that look like leftovers from the Spanish Inquisition. Made of wood, metal, leather, and rubber, and often pneumatically powered, these collectors are specifically made to harvest truckloads of seed. There are also seed hullers, meant to clean large seeds of their outer layers, that have all the elán of the contraption that pulled Charlie Chaplin from one end to the other in *Modern Times*. And never forget all those convoluted troughs made to spread various fungicides and other preparations evenly over seeds so they will never become infected while in storage. Thankfully, the small collector can fall back on his or her hands and perhaps an old kitchen strainer or two.

The time to collect seed is just before the seed ripens. Luckily, in most plants this doesn't all happen at once. For example, practically all flowers that bloom in racemes open from the bottom, allowing seed pods to form long before the last buds have opened at the top. Wild hyacinth (*Camassia scilloides*) and grape hyacinth (*Muscari bortryoides*) come to mind. But a few that bloom in racemes, like the blazing stars (*Liatris* spp.), reverse the order and bloom from the top down, yet still leave seed pods and new buds on the same stem.

Flowers that bloom in bunches, or umbels, like the flowering onions (*Allium* spp.) or the carrion flower (*Smilax herbacea*), usually all open about the same time. Later, the seeds of the first appear as shiny black beads surrounded by the brown remnants of the flowers, and in the second, the seeds are inside blue-black fruits.

Sometimes I use paper bags held beneath the seed heads, then shake the plant and allow the seeds to fall. Or you can cut off the seed heads and, later in the evening, shake them into proper containers.

When it comes to seeds found inside fruits, there is a different procedure. If you're interested in starting an orange or grapefruit tree, merely eat the container and don't swallow the pits. With other, less appetizing fruits, you can remove the seeds using your fingers, then set the seeds aside until they are thoroughly dry. Some collectors advise using a blender, mixing the fruits in water, then straining out the soft parts, leaving the seeds behind. But I never collect so many that messing up the blender is really worth the effort of cleaning it up.

In late summer and on into autumn, our kitchen table is often the center of seed-cleaning activity. The idea here is to separate the wheat from the chaff, the chaff being all those little bits of dried leaves, pods, stems, and flowers that accompany the seeds. Large seeds are easily picked out of the surrounding debris; smaller seeds can be collected by using a kitchen strainer.

For the best storage containers, use small Kraft paper bags or those little coin envelopes sold to hold stamps or coins. Either is perfect for everything except the largest seeds. Those polyethylene baggies with tight seals are never good for long-term storage. If the seed is not perfectly dry, once it is sealed inside a closed and damp atmosphere, most seed will rot. That's why paper or glassine containers are the best. Sometimes gardeners with a knowledge of origami wrap seeds in tiny packets of folded paper that look like exotic paper birds from Japan, but I've never mastered the art. Never buy those cheap white paper envelopes from discount stores; they often have tiny openings at the corners where small seeds easily fall out like pepper from a shaker.

Seed-eating insects are usually not a problem with the small collector, especially when dealing with most seeds collected in the United States. On the other hand, arboretums and large seed nurseries often have problems with insects that come in from tropical countries and can easily destroy a whole crop of seeds. Talk to anybody who works in the various research departments of the nation's botanical gardens, and you will soon learn of the dangers involved.

But that doesn't mean that being clean is not necessary in the small collection. It certainly is. Never store seeds in a dirty container, and make sure your place of storage is cool, dry, and clean. If you ever find contaminated seed, the best thing to do is throw it out. The fungicides and insecticides used to kill these invaders are not always the best things to fool with.

I've mentioned keeping records before, and now is the time to mention it again. The human brain has often been compared to a sponge, but we must remember that sponges have holes. The older you get, the more information the brain is likely to contain, and unless a particular plant name has been emblazoned in your memory, if not used, it will soon evaporate. So label everything with the name of the plant, the date of collection, where it was collected, and, if there's room, a note or two on the ultimate flower or size of the plant.

Because of the laws of genetics, when dealing with garden cultivars, you may discover that many times the seed will not produce the plant you really wanted. Yet on occasion, that particular seed will produce something entirely new. That's part of the delight in raising plants from seed.

Finally, believe it or not, ethical questions arise when collecting seed. I have a friend who was visiting the Emperor's Garden in Japan and pinched seed from a particularly beautiful Japanese magnolia—and the guards saw him do it. They left him alone but followed closely behind him for the rest of the afternoon. Today, he admits they were correct in watching him. Years ago with small touring groups, a few missing seeds would never be noticed. But today, with the hordes of tourists out there, a few seed collectors could easily denude a collection. So unless you have permission, don't collect in private gardens. I always ask the head gardener if I really hanker after a seed. On the other hand, taking small amounts of seed from a field full of goldenrod and asters should not be a problem.

If you really become involved with collecting seed, it's a good idea to pay close attention to the local, national, and international laws about the conservation of rare and endangered wild plants. There is an international convention called the Convention on International Trade in Endangered Species of Wild Fauna and Flora, shortened to CITES. Almost one hundred countries have signed up with CITES to protect all the rare plants in the world.

As to seed brought into the states from other countries, custom officials usually do not bother with small amounts of seed from a plant protected by CITES. However, do not try to bring in large amounts of seed; always declare it. Remember, live plant material comes under different rules and is a much more complicated process than importing seeds. All the seeds that I receive from England and Europe come in small paper envelopes, stamped by the customs department of the countries involved. I've never had trouble importing such seed in more than twenty-five years of gardening.

So when collecting, only take the seeds you can use. And remember, by collecting and propagating plants from seed, you are helping to practice conservation and ensure the continued existence of our wildflower and garden heritage.

INDEX

C

D

M

N

O

T